KUXAN SUUM

PATH TO THE CENTER OF THE UNIVERSE

DORA MUSIELAK

AuthorHouse™
1663 Liberty Drive
Bloomington, IN 47403
www.authorhouse.com
Phone: 1-800-839-8640

First published by AuthorHouse 9/14/2010

ISBN: 978-1-4389-5289-5 (sc)

Printed in the United States of America

This book is printed on acid-free paper.

To the dazzling stars in my universe,

My daughters Dasi and Lauren,

My husband Zdzislaw,

And the memory of my parents

Contents

Prelude of Kuxan Suum

IN ANCIENT TIMES, WHEN THE MAYAN KINGDOM enjoyed its power over soaring pyramids and stone castles with observatories raised to the heavens, there lived a princess gifted with an intellect like no other. Her name was Da'Lau, which means "one who seeks." The subjects of the empire whispered with awe the princess's name because she was smart, possessing a bright mind full of insatiable curiosity.

Da'Lau was born under the spell of the Pleiades, a twinkling cluster of stars also known by her people as Tz'ab. Swinging in her cradle the infant princess would gaze at the night sky, awestruck by the glorious beauty of the luminous stars. In the stillness of night, the babe spent hours enthralled by the splendor of the million points of light that shinned against the obscurity of the celestial vault. Ever since, the princess developed a peculiar affinity with the heavens.

One summer night when she was six, Da'Lau caught sight of tiny fireballs crossing the sky, yet fleeting, lasting an instant shorter than a single breath. She asked her teachers about the little stars falling all around, but no one knew. So the little princess imagined they were the sparks cast out by the burning of the cosmic fires.

Mystified, she also observed the Earth's Moon moving over the firmament and noted how it changed every night, from a full illuminated disk to a thin silver crescent. The young girl wondered why some nights the Moon would disappear completely from view only to reappear nights later, grow from slender to full, and start the cycle once again.

"Why does the Moon look bigger when it hangs low near the horizon," Da'Lau asked when she was ten, "than when it's overhead at night?

No one knew. So, she intended to one day explain it and the reason behind the lunar cycle. Later, she promised herself to solve the mystery of the Moon's dark blemishes, visible even by day.

K'uk', an enchanted bird with colorful plumage and a long stunning tail, patiently listened to Da'Lau's questions. This magical quetzal had befriended the princess as a child, carrying her into the air, past the puffy clouds, so she could see all there was to see of our beautiful home planet. Soaring high on the wings of the quetzal, Da'Lau discovered the Earth as it really is, a wonderland with turquoise blue oceans and dense green jungles, tall mountains and gentle rolling hills. From above she beheld unbroken plains that stretched into the distance like an emerald carpet woven with grass and rich soil. In the midst of all that, she spotted villages and great cities swarming with people that looked so small like the teeny ants in her own garden.

In those exhilarating childhood excursions with her bird friend, Da'Lau discovered towering pyramids peaking through sweltering jungles, and farther away she distinguished peculiar structures in the middle of arid deserts. K'uk' lifted the girl over majestic volcanoes crowned with white snow and lulled her over singing waterfalls hidden in the middle of rain forests; the quetzal flew her over smooth pastures and lovely fields sprinkled with colorful flowers. Since then the princess learned to cherish the Earth, as a child loves her mother.

Some nights, when she could not sleep, K'uk' took the princess beyond the land she knew. Mesmerized, Da'Lau watched the world in darkness, with parts of the globe flickering with lightning storms, forest fires. The glow of the auroras painted the polar sky with haunting colors. Secured on the wings of her companion and guide, the young girl floated through giant fronds of light and saw how the sunrise transformed the ocean into a landscape of copper and golden waves adorned with lace of white foam. And as they soared high, Da'Lau saw the sky full of hues and tints of all colors in the rainbow. Those nocturne trips upon the wings of the quetzal further widened her senses to the wondrous of nature.

Her father, an ambitious Mayan king, enjoyed his power over towering pyramids, beautiful palaces and temples, cities and villages and everything within the vast region that was his property, including the people. And so the king was keen for her to wed in order that he might extend his dominion. He ensured that young Da'Lau was educated in traditional arts, protected from the realities outside the walls of his palaces. Despite her father's prodding, the princess was not ready for marriage.

Da'Lau spent her days in the temple's library, reading the books written by ancient scholars, for she was born with a great yearning. Possessing an intrepid spirit, her mind soared beyond the limits of the world. She desired to discover the laws that govern the universe, imagining other worlds far away from the Mayan sun. Never before had a princess been so intelligent nor so thirsty for knowledge of the cosmos.

As she grew up and learned the laws of nature, the princess would meditate for hours, wondering what was beyond the confines of the Earth, the Moon, and of the Sun itself. She desired to learn what made the stars move and sought knowledge about the origin of the cosmos. Nowadays such gifted girls walk the halls of many schools in the world; but in those times they were rare and people saw them with startling curiosity. Yet, not everybody appreciated her inquisitive mind, especially the temple priests who viewed her stargazing with contempt.

The Maya believed that the Sun, the Moon, and the stars were gods and the people worshiped them with special rituals. They believed that the gods guided the heavenly bodies across the sky, continuing their journey through the underworld, threatened all the way by evil gods who wanted to stop their progress across the sky.

For this reason, and to ensure the continued survival of the universe, the Maya performed sacred rituals, self-mutilation, and human sacrifice, thinking that such acts would please and help the good supernatural beings that inhabited the heavens. For the Maya, death for the sake of their gods was a privilege—the priests assured them that such ultimate sacrifice would confer them immortality.

But Da'Lau could not imagine it. The princess speculated that if there was Hunab Ku, the supreme god and creator of the Maya, how could a lesser deity threaten the course of life on Earth and annihilate the infinite cosmos? She believed there was only one immortal omniscient god, the god of gods, the only one who created and ruled the world. For that—for her beliefs and for asking a truth no one knew—many people in the court ridiculed Da'Lau and rejected her assertions.

For the elders and the priests, Da'Lau was no longer a curious child but a rebellious young woman that questioned her people's beliefs. The temple priests rebuffed her and finally forbid her from their ceremonies, simply because she asked questions they could not answer. And she was shunned. Alienated, the princess found refuge in her books. Only her loyal friend, the enchanted quetzal, continued steadfastly by her side and listened.

One spring morning after her studies, the princess sat in her garden, pondering over what she had just learned. She wondered where the world ends, she pondered about the meaning of time, and whether time is eternal. By observing the night sky, it appeared to her that the motion of the stars and the planets was deeply interlinked with time. But she could not be sure.

Then, as it was her custom, the royal maiden asked K'uk', her bird friend.

"Where is the end of the world?"

This time, the resplendent quetzal replied that she had to fly far, far away to find it. Da'Lau knew the sky is vast, boundless, and fantastic—an eternal paradise where space and time lose all meaning. So she inquired further, eager for an answer:

"How do I go to such faraway place?"

Tilting his crested head to fix his round eye on the princess, the bird added:

"Since you long so much for that heavenly place that is not within reach for humans, you shall have the ability to fly as far as you desire, navigating through the world of eternal truths."

And in an instant a pair of white wings enveloped the graceful body of the maiden. Da'Lau was elated.

"Thank you, my little feathered friend! With these wings I will fly to the end of the world. I will go to the farthest reaches of the universe, to seek knowledge, learn the truth, and to find the heavens I long for!"

The princess was jubilant and spread her new wings made of soft down that shined with the iridescence of pearls.

PRELUDE OF KUXAN SUUM

That night, when the pale moon was a silver crescent, the princess climbed the steep stairs of the major temple. With every step her heart throbbed, striking against her young breast. But the long ascent gave the princess more courage. Upon reaching the summit, she stood before the ceremonial altar, where the eternal fire burned with golden flickering flames, raising its prayer to the heavens.

Bathed in starlight, the princess removed her golden headdress festooned with emerald colored feathers and multihued precious stones. She removed her royal pendants and bracelets encrusted with jade and lapis lazuli and deposited them in the sacrificial urn. Her only adorn now was her long black hair that fell over her winged back like a cascade of glassy obsidian.

After praying, Da'Lau lifted her head skywards, spread her pearly wings, and flew in search of something she could only find among the stars. Her mystic quest began, her search for the causes of the cosmos and the principles of knowledge and wisdom.

Da'Lau slowly ascended, defying gravity, floating effortlessly through the air. She left below her royal home and the sacred temples illuminated by torches. Flying higher, she could see the rapidly receding outline of the brownish green continents surrounded by the serene azure oceans. Spiraling around the Earth through an unseen path to reach the heavens, Da'Lau flew until she found the end of the atmosphere, beyond which not even the magical bird could fly. The princess contemplated for the first time the full Earth rotating below her, a cerulean sphere embraced by a tenuous layer of protective gases.

Riveted, looking down on the planet, she saw every sunrise and sunset of the world, every one! Earth seemed like a jewel speckled with swirling white clouds, a sphere spinning in the midst of black space. Da'Lau was profoundly aware of the wondrous beauty of the earthly landscapes, remembering the awesome scenery she had admired as a child.

"That *is* Kab', my home!" she cried out before soaring higher through the dark sky with her wings shimmering under the moonlight.

☙ ☆ ☆ ☆ ☆ ☆ ❧

At a far distance away all was quiet, and the princess felt a pang of loneliness. All about her was darkness so intense she could almost reach out and touch it. After a few moments, she overcame her apprehension and kept on her journey, thinking that even the endless sky must end, though she could not imagine it.

In her trajectory, Da'Lau first went by the Earth's Moon and discovered it was another world, a spinning rocky sphere but somehow not like ours. Astonished, she scanned the airless desolate *Luna* and hovered over it, surveying its dusty valleys and gigantic craters. She beheld the bright highlands and the darker smooth maria. At last she knew what the curious dark spots on the Moon were and smiled; she had solved the riddle!

Leaving the Moon's neighborhood the darkness of deep space overtook Da'Lau. Her heart fluttered like a butterfly when she turned her head around and saw how infinitely small the Earth is, how vulnerable that azure-hued dot with white specks is, almost imperceptible amidst the blackness of space. And with that as her last glance, Da'Lau bid adieu to her beloved planet.

The brave princess then searched for Venus, the brightest point of light known as the Great Star much revered by the Maya. The girl knew that it was not a star but a planet about the size of Earth that orbits around the Sun. As she flew over Venus, she plunged her gaze toward the cloudy orb, probing the peculiar incandescent clouds, striving to penetrate the hidden surface below. Venus—she discovered—was a scorching world, enclosed with acid vapors and a choking atmosphere that would be asphyxiating for people. Intrigued she wondered why the morning star planet so beautiful seen from the Earth was so deceptively hostile once near.

The princess flew onward through the sweltering space and stopped over the closest planet to the Sun, the one that looked so old with its heavily cratered surface, engulfed by a complex plasma nebula. Still floating over Mercury, Da'Lau caught her first sight of the splendid Sun, our mother star. From afar she watched the luminous ball of plasma, colossal compared with the planets, rotating with violent explosions on its atmosphere that blazed with mighty whirlpools of fire. Incandescent solar flares forcefully extended to reach far into the vast space between the distant planets.

"Oh mother star of the Earth, how beautiful and powerful you are, and though I know you love me and give me sustenance, you won't allow me to get near you."

PRELUDE OF KUXAN SUUM

Changing course, the princess continued her flight across the immense space that separates the planets, with an invisible solar breeze gently pushing beneath her wings.

"I will go just a little farther," Da'Lau said to herself, intrigued by the mystery of interplanetary space.

First she went by a rocky red planet with the largest mountains she had ever seen; its crimson surface was covered with craters, plains, and colossal canyons. The soil seemed lined with signs of ancient rivers now dried. The uninhabited planet was a world so much like hers! It had polar ice caps and clouds in its atmosphere. But it was very cold, and the desert ground loomed lonely under the pinkish sky. Something in that planet was familiar but she knew very well she had never seen it.

Because of its color the princess concluded the desolated planet was the reddish point of light in the sky her teacher had taught her to track many years before, as it became visible in the morning after its period of invisibility. The tutor had trained her to measure the times when the planet's motion reversed its direction relative to the stars. The girl smiled with pleasure remembering her scholarly lessons.

She left the Red Planet and crossed paths with glittering comets as they moved on their periodic voyages, with their spectacular icy heads and long sinuous tails shining against the light of the Sun.

In the quietness of the obscure night she went past giant gas planets that possessed many moons dancing around them, rotating among the gleaming colored rings of rocky dust that made the orbs look very pretty. The biggest icy planet had an enormous reddish cloud of rising gases that, like a giant shell, revolved in a rhythmic cadence. What a sight that was!

The bitterly cold darkness of the cosmic sea enveloped Da'Lau and made her shiver. And when she thought that was the end of all, suddenly she discovered a strange object, too small to be a planet yet bound to the Sun like the others. It seemed like a lost child at the outskirts of the solar system but the tiny planetoid had moons circling around it, as if guarding it.

Soon the Mayan princess was on the verge of the stars themselves. She finally had come upon the threshold that had to cross all those who wish to enter into the realm of truth, knowledge, and wisdom. Da'Lau became herself a beam of light so glorious, so radiant, like the light that has traveled across the cosmos from the beginning of time. In an instant past and future were interchanged, space and time became one.

In the distance, thousands and thousands of lights shined. Da'Lau was dazzled as she spotted a kaleidoscope of resplendent stars, brighter than the most luminous gemstones in her feathered tiara. There were stars enveloped in gold and crimson auras, some spewing gases bluer than the turquoise of the ocean.

Continuing her interstellar voyage the princess passed near Proxima Centauri, the star closest to our Solar System but one that Da'Lau had never seen because the red dwarf star is too faint to be seen from Earth. Marveled at the beauty before her, made the princess curious to reach other parts of the Milky Way, our galaxy called the Wakah Chan by the Maya.

Flying through the immense vacuum distance that separates one star from another, the tranquility of eternal night was soothing to the young girl. But close to the stars themselves, giant turbulent currents of searing gases engulfed all matter. Da'Lau witnessed the violent displays of stars dying, throwing multicolored gases like streamers radiating hotter than fire. Near her route she found a white dwarf, an exhausted star enveloped in its glowing cocoon of gases. The star was ending its life by casting off its outer layers of colored gases.

Farther away the princess noticed many more stars that had already died. She wept mournful tears at the sight of the stellar graveyard, remembering her own Sun that one day would also die and her people would perish with it. But it was consoling to know that one day the Sun of the Earth would burn out and shroud itself with stellar debris before extinguishing its light forever, but not for another five thousand million years.

Da'Lau continued her voyage across the universe, seeing what mere mortals could not. Her body was a glimmer of light as the stars rose in jagged flashes, guiding her across ethereal paths known only to those who seek wisdom. The girl's eyes shone, reflecting the radiance of giant stars whose immaculate light has traveled the vast distances of the cosmos to illuminate the paths of every man and woman who has ever lived. As the princess flew among the dazzling stars, a sublime breeze hummed a tender lullaby, a song so pure and enchanting, a serenade to the heavens. Da'Lau flew protected by the stardust of that paradise.

The princess traveled at the speed of light, and yet her voyage was long. But her desire to learn and discover kept her from falling into the shadowy recesses between the bright and dark stellar matter. In the cosmic void, her sense of spacetime was truly high dimensional.

As the princess continued on her long journey, fifty-five hundred light-years away, she came upon the Omega Nebula, a stellar nursery sculpted by colorful stellar winds, undulating shapes of dense cold gas and glowing dust, illuminated by baby stars, and the swirling auras that enveloped grown stars. It was the most magical sight for the princess!

In her long journey, Da'Lau traversed other solar systems abounding with fluffy planets and strange moons, so different from our own. She saw stunning worlds of haunting, alien beauty. Bewildered, she spotted double star systems and giant planets circling other suns and was amazed.

"Are those planets like our Earth?" she wondered, "Is there life in those worlds similar to ours?" From that vantage view she could not tell.

She kept on her cosmic crossing, awestruck with images of billowing clouds of dust ablaze with the light of newborn stars. The princess was seeing the Mountains of Creation in a region of deep space some seven thousand light-years away from Earth! Da'Lau wept in ecstasy, for she had never seen such colorful splendor. And as the princess flew through the vastness of space, her tears were spread by the astral wind to form more glowing stars to guide her journey.

Leaving the Milky Way, Da'Lau crossed nebulae, large glowing colored clouds of interstellar gas and dust. From that cosmic vantage Da'Lau had a glimpse of her home galaxy, majestic and splendorous with millions of stars in the center and spiral lanes of dark dust and iridescent pink clouds in which new stars were forming, all shimmering against the dark matter. It was so beautiful out there that the princess cried of sheer happiness and wonder.

"I'll just go a little farther," she told herself, curious to discover that there were many more clusters of million stars in all directions.

As she approached Andromeda, the magnificent giant whirling galaxy greeted her head-on, appearing to move in a collision course with Wakah Chan. But Da'Lau did not stop. She continued her spiritual search and journeyed past glittery starburst galaxies and glowing stars huddled in peculiar shapes that appeared like swarms of fireflies in a summer night. She contemplated stars being born in rapid whirls of gas and dust, and other stars dying with spectacular explosions that illuminated the dark cosmos.

On the curled arms of a faraway galaxy, Da'Lau spotted a most spectacular nebula, sculpted by the action of torrential winds and scorching radiation from monster stars that inhabited what looked like an inferno. One such star about to erupt was surrounded by two billowing lobes of gas and dust, while a hurricane blast of the stellar wind seemed to reach out to infinity. Space quivered around her, ripples of invisible energy seemed to contract and expand the boundless void. In that instant, order and chaos were dynamically and mysteriously intertwined.

Enthralled by giant cosmic flashes, explosions at the centers of galaxies from refulgent objects called quasars, Da'Lau witnessed a galaxy plundering a group of stars, like a thief in the dark; she saw galaxies running towards and away from each other, accelerated by a mysterious force that no one understands yet causes the universe to expand. In the farthest region of space she detected the most titanic explosion she had ever seen—a fulgent burst of energy appearing to originate in a distant galaxy. The vacuum of space insulated the roar of the blast. It lasted a mere few seconds, but the bright afterglow remained, illuminating her transcendental voyage.

The mysteries of the cosmos unfolded before Da'Lau's eyes. Traveling so far into infinity, she had a vision of heaven. Her soul leaped into another dimension. She was transported into an ethereal region of light and there she gazed at the face of a divine being, the God of everything. Da'Lau saw the Immortal Creator in the exquisite beauty of the universe. And as the Mayan princess went deeper through shimmering nebula into the magnificent cosmos, the celestial breeze whispered tenderly: "Kuxan Suum, Kuxan Suum!"

☙ ☆ ☆ ☆ ☆ ☆ ❧

Meanwhile in the palace, at dawn, the imperial guards sounded the conch shell horns when the alarmed king discovered that his beloved daughter was not in her royal chamber. Her feathered headdress was found burned to ashes in the sacred urn, and her jewelry was smoldering amidst the black soot. Among the remains the father found her childhood treasures, the sacred coral seeds and diminutive gemstones that the princess carried in a small leather pouch tied to her waist. Stricken with worry, the monarch ordered every subject in the Mayan realm, young and old, to look for Da'Lau. They searched every corner of the palaces, in the gardens, and in the chambers of the sacred temples, but nobody could find her.

The king was heartbroken. He cried bitter tears, for he knew he was to blame for the disappearance of the princess. He did not know how to protect her from the rejection of her own people; he, like everybody else, had not understood nor valued her uniqueness. During Da'Lau's childhood, the Mayan lord had dismissed her desire to learn and to discover truths that transcended life itself. Later on, troubled by her scholarly inquiries, the father had locked her up. The king had seen Da'Lau only as a precious jewel to be traded to expand his monarchy. Now the sad man regretted all that and began to see his daughter as she was, a special being born to seek knowledge. But at that moment, the king could not imagine where she was or Da'Lau's true destiny.

The Mayan ruler was overcome with melancholy and grief. After a desperate vigil, he ordered the royal guards to proclaim a hefty reward for anyone who'd find the princess. The king offered a thousand gold pieces and a chest full of the venerated jade stones, obsidian, turquoises, and other exquisite jewels of immensurable value to anyone who'd bring Da'Lau home.

Young princes and noble gentlemen came from other kingdoms and distant lands to search for the princess. For them the reward was tempting and more so as the maiden was considered for marriage. Dismissing her strong intellect, the young men who had seen her before were at once enchanted by Da'Lau's long black hair beholding a glassy luster like obsidian. Her large dark eyes had pierced their souls with a penetrating gaze they could not forget. But now, no one really knew where to look. After fruitless searches, all gave her up for lost and left.

At the same time, the king had sought the help of a sage scribe, the astronomer who spent days and nights in the highest pyramid tower, observing and recording the complex motion of the heavenly bodies. Yaxk'in, his name was, had taught the princess the constellations in the sky when she was little. Having been her teacher he knew of the yearning and deep thinking of the royal pupil, for mere legends were not enough to satiate her thirst for knowing.

The astronomer had understood that since the first hour the princess opened her eyes to the exquisite enchantment of the blue sky above her, Da'Lau had developed an intimate connection with the Universe. As she grew up into a young woman, Yaxk'in was captivated by Da'Lau's intellect and fine mind, comparable to his and that of the great philosophers and scholars. He taught her mathematics so she could combine them with her celestial observations and understand the motion of stars and planets.

The tearful king had come to him pleading.

"O great observer of the heavens, can you find my beloved daughter? The gods must have taken her into the depths of the darkness!"

The astronomer doubted that.

"The gods did not take the princess, mortals did!" Yaxk'in blurted out.

And even without the monarch's beseeching, the teacher intended to find his pupil.

That afternoon, after all searches were exhausted and the king had resigned to the loss, the Mayan astronomer left his observatory post and sat under the Ceiba tree, there on the same spot where the princess used to read.

"Where did they take her?—the good man said to himself— "I would go to the end of the world to rescue her if she were there!"

Nearby, a quetzal of colorful plumage was perched on a tree branch, the same bird who conversed with the princess every morning. After listening for a while, K'uk' felt pity and told him:

"Our beloved princess has flown far away; you will find her among the stars!"

"How am I going to find her in such immense cosmos?" Yaxk'in cried out utterly frustrated, fully aware of the impossibility of such mission.

But the resplendent quetzal had disappeared, leaving the perplexed man to his lamentations. He struggled with the idea of a cosmic journey, which seemed unfeasible and utterly impossible, but his desire to find the princess and his belief in a higher power beyond human comprehension sustained him.

As the blazing Sun set below the horizon and the first stars appeared in the clear sky, Yaxk'in returned to the observatory to seek guidance from Hunab Ku, the supreme god of creation. After meditating, the astronomer proclaimed, lifting his eyes to the heavens:

PRELUDE OF KUXAN SUUM

"If I could travel fast
At the speed of light,
My body would not be,
But my soul
And what I feel
Would fly to her
To never leave."

As the last of his words were uttered, Yaxk'in noted a fireball blazing forth against the dark background of space with a splendor that outshined every star. The bright object in the sky had a curved tail that looked like a cosmic finger pointing down at an unseen spot. It was a comet like none he had ever seen and he wondered if that was a sign from the gods. He watched the comet throughout the night until his body collapsed with the heaviness of sleep.

Early the next morning, long before the first light of dawn glimmered on the horizon, the astronomer opened his eyes after a restless slumber and discovered he had wings, long, feathered limbs that would allow him to fly like a firebird. Not wasting a moment, Yaxk'in took to the sky in search of Da'Lau.

And just like the princess, the good scholar went past the Moon and the pulsating yellow Sun, where the blistering solar flares lashed with torrential energy, keeping him at bay. He flew by the giant planets but did not stop, for the numinous quetzal had said that the princess was star-bound and had flown far, leaving behind many suns.

As Yaxk'in left the realm of the planets circling the Sun, and found himself in the darkness of the space beyond, he inspected the sky, rippling with millions of suns, each one shrouded by astral winds with lucent highlights of exquisite beauty and divine splendor. He recognized blue giant stars in one direction, red dwarfs in the other. Flashes of dazzling colors illuminated the void of space.

Yaxk'in discovered that the Milky Way galaxy was littered with stellar relics, planetary nebulae of chaotic structure with hot white dwarfs at the center, rich in clouds of dust, some of which formed long, dark streaks pointing away from the incandescent stars. In other regions, he saw scarlet clouds like rubies, others of deepest vibrant blue like sapphires, and other nebula of many colors adorned with golden rims. The otherworldly clouds moved rapidly, propelled by a strange kind of energy.

The panorama before his eyes was awesome, but his heart was full of trepidation. Yaxk'in had told the princess the ancient stories of the origins of the cosmos, and when she was little he had entertained her with legends of gods that resided among the brilliant stars. But the reality before him was not as it was told nor how he had imagined it. He was astonished! Hundreds of galaxies populated the heavens and moved at great speeds away from one another. Despite his effort, he could not distinguish the outer edge of space. The amazed Yaxk'in discovered in an instant that infinity was truly in every direction!

When his courage dwindled, a beam of remote light beckoned him with a periodic pulsing rhythm, as if it were a lighthouse promising safe harbor. The extraordinary light beamed right into his heart. With renewed hopefulness, Yaxk'in kept on his search, lulled by the murmur of galactic background.

And like Da'Lau, the astronomer traversed many galaxies, each galaxy containing millions of suns. And like her, he discovered that stars are born and then die, just like human beings do. He recognized planets circling red dwarf stars and was dumbstruck by supernovae, the colorful explosions caused when massive stars die, exhausting their fuel and collapsing. Hypnotized, Yaxk'in saw geysers of color spewing from the cores of active stars. Wild vortices of gas and burly currents of unseen matter shook him wildly.

Yaxk'in crossed a spectacular nebula divided into parts by dark, obscuring dust lanes and filaments of luminous gas. He discovered in the cosmic cloud a stellar nursery where young and embryonic stars were hidden in natal dust and other matter. Moving on, the astronomer smiled lovingly and tenderly at the sight of a newborn star enveloped in its cocoon of glistening scarlet interstellar gas. In other regions of space, some stars seemed to be scattered randomly like discarded jewels in the cosmic abyss between galaxies.

After traveling far into the dark cosmos, Yaxk'in glimpsed a peculiar band of stars, like a highway in the sky studded with sparkling diamonds. The astronomer wondered what it was, and the celestial wind murmured in his ear:

"*It is a road leading to the center of the universe; follow it!*"

He was stunned, for he did not know if this winding path in the sky would lead him to the princess. But the glorious beauty of the stars reminded him of her eyes, and he knew in his heart Da'Lau must have wandered on this lovely lane, paved with millions of colorful stars. And he perceived in the cosmic distance ahead of that path the glorious splendor of eternal light that beckoned him.

☙ ☆ ☆ ☆ ☆ ☆ ❧

It was never known if the astronomer ever found the Mayan princess. But if both passed through the same incorporeal path and finally met at the center of the cosmos, it would be impossible for them to go back to her kingdom. For, you see, in the center of the universe there is an infinite large and massive black hole. If one crosses past its inner boundary, what people now call the event horizon, it is impossible to escape, not even flying at the speed the light!

And since that night when the princess became one with the stars, the path in the sky leading to the center of the universe became known throughout the Mayan world as Kuxan Suum.

And now you know the story of the princess Da'Lau who will live forever, as stellar dust and as cosmic light traveling across our infinite universe …

Preface

THE TALE OF THE MAYAN PRINCESS wandering across the Universe is, of course, a fantasy. However, her idealized intergalactic voyage in spacetime helps us to frame a vision of the cosmos in all its splendorous beauty. It also helps us to ponder and ask questions related to what we cannot see and the exploration of space we have yet to accomplish.

Kuxan Suum is a metaphor. My intention in writing such allegory was to bring to the forefront several scientific challenges faced by engineers and scientists involved in the space exploration program. In writing a fantasy, telling a story of a human being traveling through deep space, I also wished to convey the sense of awe for the human feats that characterize the exploration of the Universe. The thread that stitches *Kuxan Suum* is its surreal invocation of travel through the cosmos—the sublime, the practical, and the science.

Over the past years, I've become increasingly interested in astrophysics, as seen from my aerospace engineer perspective. Specializing in space propulsion was a choice I made to get closer to my childhood dream of finding a way to voyage to the stars. But how could I conceive a rocketship without knowing the space environment where it would travel? Not only does a rocket engineer need to know and appreciate the enormous distances to be traveled, but she must also understand the physical challenges of a vehicle moving in outer space.

Like the Mayan princess, I've envisioned traveling from planet to planet. As a little girl, I imagined places hidden in the sky and often wondered about remote worlds in the Galaxy. More than anything I wished I had the power to navigate among the stars. Later, after learning basic astronomy, I asked myself, *how can anybody go fast enough to make a voyage across the immense cosmos*? In the fantasy, the princess Da'Lau moves at the speed of light, and as such it makes her to be a beam of light, a photon. But even in that form the whimsy remains, as even a beam of light could not traverse the Galaxy within a human lifetime.

Of course, I could have written that Da'Lau traversed deep space using faster than light (FTL) travel. That would have implied that she used spacetime warps, wormholes, dark energy, or some other unknown form of FTL. But in the end, I decided to let the reader speculate and perhaps assume that somehow the princess found a way to interact with spacetime itself to cross the cosmos within the scope of the fantasy, just to show us how extraordinary and spectacular *our* Universe is.

The indomitable beauty of the sky and my fascination with space exploration persuaded me enough to encourage me to write this book. Some time ago, while giving a course in propulsion, I realized that aerospace engineering students needed to learn not only how to design rockets but also need to know a little bit about space sciences, topics absent in the curriculum. Because, to appreciate the challenges of spaceflight beyond low Earth orbit, we need a virtual tour of the cosmos, to imagine and become aware of its complexity and size in order to conceive such a voyage. We also have to establish the enormous distances that separate us human beings from other bodies in the sky. Further, we must have a sense of the complexity of the space environment to help us to characterize and gauge the cosmic paths our spaceships would navigate.

These imperative premises and my enthusiasm for the exploration of the Universe brought me to conceive this book, linking modern topics of Astronomy, Astrophysics and Astronautics.

I decided to open with the *Prelude of Kuxan Suum* because this fantasy combines the scientific ideas needed to better understand space exploration and the challenges of a journey to the stars. The fable aims to inspire young readers and communicate some of the wonders in the sky. In each chapter I give a sketch of the concepts interwoven in the Prelude of Kuxan Suum, ideas containing the scientific truth behind the tale. Part of that vision includes a description of cosmic objects such as black holes and exoplanets, and a synopsis of fascinating concepts such as the nature of the time and the origin of the Universe. I will explain these topics in simple terms since a complete and comprehensive treatment would require large volumes and a rigorous treatment of mathematics and Astrophysics. This is why I will give just sufficient details to provoke others ideas and encourage the reader to investigate further.

PREFACE

For me, the age of space exploration is an era of exciting discovery. Much has been achieved since it started—less than sixty years ago—and much more will be realized in the future. The development of space rocket technology has allowed us to build sophisticated space probes that have extended our reach to the outer edges of the Solar System and beyond. I would be pleased if every reader becomes enthralled by the mysteries and secrets of the Universe, and gets an idea of the thrilling feats that characterize the exploration of space. I hope that this book will inspire them and, after reading it, they begin to look at the sky in a different way.

It would be impossible to address the subject of the exploration of space without mentioning the vehicles that launch the satellites, robot probes, and space observatories to Earth orbits, and also that transport astronauts to the International Space Station (ISS). I will only mention the manned space launch system of the United States known around the globe as the Space Shuttle, or more properly NASA's Space Transportation System (STS). And since the entire Space Shuttle system includes the Orbiter (the reusable vehicle which carries the crew and lands like an airplane), when appropriate I will refer to the Space Shuttle Orbiter.

In referring to space I mean the environment beyond the Earth's atmosphere, in fact beyond the Kármán line, which is 100 kilometers (62 miles) above the surface of the Earth. With the exception of flights to the Moon in the twentieth century, most of the manned space flights have been to low Earth orbit, including transport to space stations. Low Earth orbit (OTB) is between 200 and 1200 km above sea level. The International Space Station (ISS) moves in an orbit that is at an altitude between 319 km and 347 km above the ground. Space telescopes are in higher Earth orbits.

I will also use scientific notation throughout the book because it is the concise manner to represent numbers using powers of ten. This notation is used to easily express large numbers such as those that we need to express the distance between the planets or the mass of the stars. Examples of powers of ten are provided in the *Glossary*.

My intention is to encourage scientific research among youngsters, as this is a fascinating field that requires much learning. Thus, I started by raising some of the questions I had when I was in school, and others that I hope to answer one day. The *Questions to Ponder* at the end of the book are intended to stimulate the imagination of young readers, because they will be the space explorers of tomorrow. I would be pleased if at least one reader answers those questions and articulates others or takes the study much further. My desire is to stimulate interest in science, mathematics or engineering careers. And of course, I also want to reach readers interested in space exploration and the wonders of the Universe.

Finally, I wish to express my deepest appreciation to NASA and the Hubble Team for permission to use the gorgeous images to illustrate this book. In 2010, the Hubble celebrated a major milestone. The most extraordinary space telescope in the history of humankind was launched from the Space Shuttle Discovery on April 25, 1990. Since then, the Hubble has traveled 2.8 billion miles, capturing an extraordinary 570,000 images of our magnificent Universe, giving us a glimpse of the striking beauty of space never seen before. One usually gives presents to the one celebrating a birthday; in this case is the Hubble that gives us the best gift of all. Happy Birthday, Hubble, and thank you!

1
How big is the Universe?

Too many to count, too far to reach,
the cosmic lights that twinkle and whisper the truth of my being…

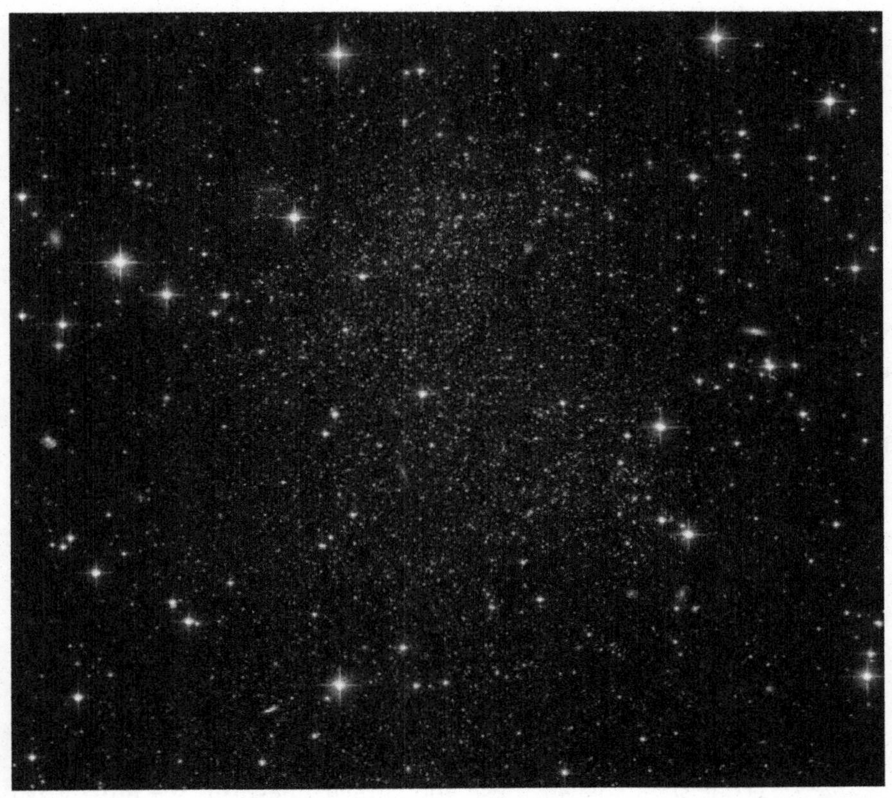

Sagittarius Dwarf Galaxy
Credit: NASA, ESA, and The Hubble Heritage Team (STScI/AURA) - Acknowledgement: Y. Momany (Univ. de Padua).

If we could observe at once—as Da'Lau did—everything in the Universe, we would notice that everything is in motion.

- ★ The Earth rotates on its axis and revolves around the Sun (our own mother star).
- ★ The Earth has its own natural satellite we call Moon, which rotates and circles around us.
- ★ There are other seven planets that circle the Sun in distinct orbital paths, some with their own moons and multicolored rings of clumped particles. In our Solar System there is at least one dwarf planet, hundreds of thousands asteroids, and millions of small bodies circling and gravitationally bound to the Sun.
- ★ The Sun (and everything gravitationally bound to it) revolves around the center of the Milky Way (our own Galaxy).
- ★ Millions of other stars also rotate around the super-massive black hole at the center of our Galaxy.
- ★ The Milky Way, a conglomerate of stars in the shape of a spiral, is just one galaxy among other billions of clusters of stars.

And that's not all. Black holes (product of the compression of stars that are ten or more massive than the Sun), neutron stars, extrasolar planets, dwarf stars (stars that when ending their activity are compressed to a size similar to Earth's size), dead stars, quasars, pulsars (neutron stars that, by their rapid spin, their radiation is perceived as very short pulses), and an array of exotics particles inhabit the heavens. Most strikingly, however, is the fact that all that is just a very small fraction of the stuff that comprises the cosmos. Yes, only about 4 or 5 percent of the Universe is made of ordinary matter. The rest is matter and energy that defies description because it's invisible and has yet to be fully defined.

That's the space environment in a nutshell. Later on I will tell you more about it. For now, let's ask: *How feasible are interstellar and intergalactic voyages*? What I ask is, *Can we travel from star to star, from galaxy to galaxy*? To help us answer these and similar questions we must learn how far the stars are in order to know how large the distances to be traveled.

The Universe is unimaginably huge and breathtakingly beautiful. For thousands of years people knew the sky is immense, but not until the twentieth century we began to grasp its real immensity. The first attempt to quantify the size of the heavens was made by the Alexandrian astronomer Aristarchus of Samus in the third century. Of course the "universe" at that time was believed to be just a tiny fraction of the actual Universe. Aristarchus (ca. 310-230 B.C.) deduced—but inadequately calculated—the distance from the Earth to the Sun. He estimated the Sun was just about 8 million kilometers (~5 million miles) away.

In reality, the Sun is about 150 million kilometers from the Earth (92 million miles). Other ancient astronomers conceived a world whose form was thought to be a sphere a little bit bigger than Earth's orbit. The thinkers of antiquity believed that the Universe was a huge sphere to which the stars were fixed. In its interior, the planets and the Sun occupied spheres of smaller and smaller radius and the Earth was set permanently in the center. That was the geocentric view of the Universe that was accepted since the time of the classical Greek philosophers and until the great scientific revolution started by Copernicus and popularized by Galileo in the 16th and 17th centuries. At that time some astronomers still thought that the entire vault of heaven, including all the stars, was a sphere about 300 million kilometers in diameter.

However, even though the contribution of those great thinkers led to an exact knowledge of the structure of our Solar System, the ideas about the structure and dimensions of the entire Universe remained relatively limited and confused until the birth of astrophysics, at the beginning of the 20th century. Now we know that the Sun, with its entourage of planets that accompany it, is just one of the many stars in our Galaxy, and that this, in turn, is just one more among the many systems of stars in which matter is concentrated.

The Universe is gigantic. Even in the minute region of the Galaxy, where our Solar System resides, the distances between planets are so vast that they are difficult to comprehend. Mars for example, our neighbor planet and future destination for human exploration, when it is at favorable opposition (the closest), it is about 56 million kilometers away. At its greatest distance Mars is over 399 million kilometers far. The two outer planets in the Solar System are so far that before the telescope nobody knew they were there. Uranus, the giant icy orb, was discovered in 1781—over 170 years after Galileo pointed his telescope to the sky for the first time. Uranus is at an average 2,870,990,000 km from the Sun —that's almost 2 billion miles away! And Neptune, discovered in 1845, is much farther, approximately 4.4 billion km away.

That's not far really, compared with the distance to tiny Pluto in the frigid edge of the Solar System. Pluto is at an average distance of 5,763,920,000 kilometers from the Earth, so far away from us that it is not visible in the sky. Because its orbit is very elliptical, the closest distance is about 4,436,820,000 km, and the farthest is 7,375,930,000 km. Pluto, now considered a dwarf planet, was discovered in 1930, and its largest moon, Charon, was discovered in 1978. Two more moons were imaged by the Hubble Space Telescope in 2005, leaving scientists to wonder how it could be. Today, we still don't know how Pluto looks like. We must wait until NASA's space probe New Horizons arrive to that distant region en 2015 to learn more about Pluto and Charon.

Let's put those distances in perspective. The Sun is almost 150 million kilometers from Earth, or more precisely, 1.495978706×10^8 km (written in scientific notation), a distance known as Astronomical Unit (AU). The Astronomical Unit is more convenient to use than kilometers when establishing the large distances within our Solar System. And so, for example, we say that Pluto is about 39.5 AU from the Sun, or almost 40 times farther than the Earth is.

Beyond the orbit of Pluto there is a region known as the heliosphere, a tear-shaped bubble in space that contains our Solar System. Its outer edge is the heliopause, the theoretical boundary where the solar wind's strength is no longer great enough to push back the winds of the surrounding stars. This boundary is somewhere between 85 and 120 AUs from the Sun (13.5 billion km away). The dimension of the Solar System can be inferred from such distance.

After reaching the edge of the Solar System, there is a very large region of empty cold space before we find the stars. Proxima Centauri, the closest star to the Solar System, is about 3.8×10^{13} km away from us. That is about 25,401 AU, or 38,000,000,000,000 km, if you prefer familiar units of distance. But after a few million kilometers, it is hard to conceptualize how far objects are in the sky, and thus kilometers are inadequate to give the distance of the celestial objects.

Consider the following analogy: If the Sun were the size of a small marble, the distance from the Sun to the Earth would be a little over 1 meter, and the Earth would be barely thicker than a sheet of paper. On this scale, the closest neighboring star would be about 338 kilometers away, and the next star would be twice as far, with nothing in between, just the vacuum of deep space! This analogy should give us an idea how far the stars really are from us, and how vast the distances among them.

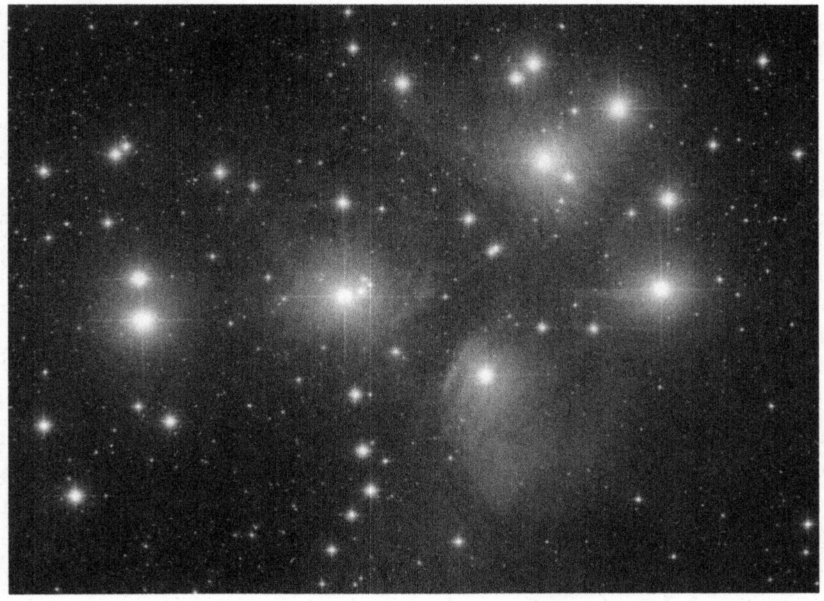

Pleiades Star Cluster. Atlas along with "wife" Pleione and their "daughters," the seven sisters, are the principal stars of the Pleiades that are visible to the unaided eye, although there are actually thousands of stars in the cluster. Atlas is between 434 and 446 light-years from Earth. Credit: NASA/ESA and AURA/Caltech.

To measure the long distances to the stars, we use a unit called light-year (ly). A light-year measures the distance light travels in vacuum in one year. Since light travels at 299,792,458 meters per second in the vacuum of space, we estimate that in a year ($\sim 3.1536 \times 10^7$ seconds) light can travel a distance of almost 9,460,730,472,581 km. That is roughly 10 trillion kilometers! Thus, we approximate a light-year as equal to 9.46×10^{12} km. Using this measure, we say that our nearest star (beyond the Sun) is 4.22 light-years away. Sirius, the brightest star in the sky after the Sun, is twice as far, 8.6 light-years away. More astonishing is the total size of the Milky Way; it has a diameter of about 100,000 light-years. And ours is just an average-sized galaxy!

Another measure of distance used in astronomy is the parsec (pc), which represents the parallax of one arc-second. You may say that an object that is 1 pc away has a parallax angle of one second of arc. And since one pc is equal to $3.08568025 \times 10^{16}$ m, that object would be over 30 trillion kilometers away. Assuming the angle is small, the distance to an object (measured in parsecs) is the reciprocal of the parallax (measured in arcseconds), that is d(pc) = 1/p(arcsec).

For example, the distance to Proxima Centauri, which has a parallax of 0.772 arcsec, is 1/0.772 = 1.29 parsec. The Milky Way has a radius of about 15 kpc (kilo-parsecs), and the Sun is about 10 kpc from the center. Therefore, the distance from our Sun to the center of the Milky Way is about 2,061,612,202.7 AU, which corresponds to about 1.91639×10^{17} miles. Try to imagine that! Well, I prefer use light-year as the unit of distance to the stars, and so I use the conversion 1 pc = 3.26 ly.

However, it is time, not distance, what limits our view of the Universe. When we look at the bodies in the sky we say we are looking into the past because light takes time to reach us. Even light reflected from the Moon, which is located about 382,500 km from the Earth, takes a little more than a second to make it across. If the Sun would explode right now, we would not know about it for eight minutes because that is how long it would take for the light of the blast to reach us.

When we see the stars, we see them as they were in the distant past. For example, if we live in the Northern Hemisphere, the smudge we see in the constellation Andromeda is actually the enormous Andromeda Galaxy, located some 2.2 million light-years away. If we are in the Southern Hemisphere, we may see the Magellanic Clouds as they existed half a million years ago. And if we would see what seemed like a new extremely bright point that disappeared in a few weeks, maybe that could be a supernova, a star that exploded millions of years ago in a far away galaxy.

Turn your eyes to the night sky, and you will see many stars perhaps not realizing that there are millions and millions, many more stars than we could see with the naked eye. For centuries people believed the Milky Way was the only galaxy in the sky. In fact, up to the beginning of the twentieth century, most people thought the Milky Way was everything! In 1924, American astronomer Edwin Hubble presented the first observational evidence that other galaxies lie far beyond the Milky Way. Now we know there are roughly 200 billion galaxies in the observable Universe.

Astronomers estimate that there are between 100 billion stars in the Milky Way alone and up to 400 billion ($1\text{-}4 \times 10^{11}$ in scientific notation). So, if just one galaxy has that many stars, and there are billions more galaxies, perhaps between 10^{11} and 10^{12} galaxies, can you guess how many stars are there? Using a simple approximation we can say that there are many trillions—probably between 10^{22} and 10^{24} stars. The number 10^{24} is big. Write it down without the help of scientific notation; this number consists of a 1 followed by twenty-four zeros and it looks like this:

1,000,000,000,000,000,000,000,000.

Such large number is beyond what is meaningful to us, because the human mind cannot comprehend the amazing size of our Universe and so the only real way to comprehend it is to compare it to something we know, especially at the elementary level. And thus, according to some grade school teachers, the number of stars is many times more than the number of grains of sand on all the beaches in the world. That would be a good analogy if we can determine the number of grains. We might count them by measuring the surface area of a beach, and determining the average depth of the sand layer. If we count the number of grains in a small representative volume of sand, by multiplication we can estimate the number of grains on the whole beach. Now you have to know the surface area of all beaches in the world. What do you think? For me, the number 10^{24} is just mind bogglingly huge and, even with the analogy, is hard to grasp how many stars are.

This number of stars is a rough estimate, of course, as we do not count stars individually. Scientists determine the mass distribution of stars in the galaxy. They also know the amount of light emitted by each type of star. So, by measuring the total amount of light in the galaxy (called luminosity), and knowing the mass, astronomers can estimate the number of stars that are there in a galaxy.

Furthermore, knowing how fast stars form can bring more certainty to calculations. The European space observatory Herschel, for example, will also chart the formation rate of stars throughout cosmic history. If astronomers estimate the rate at which stars have formed, they will be able to estimate how many stars there are in the Universe today.

But whether there are 10^{23} or 10^{24} stars, the fact remains the number of cosmic objects in the sky is colossal!

The Universe is indeed stupendously humongous and stunningly peculiar, and it makes one wonder, *how did all begin? Where the stars, the Moon and everything we see in the sky come from? How did we come to be*? Let's then go back to the beginning of everything.

2
A Universe is born!

A very long time ago, our world did not exist.
Over 14,000 million years ago, nothing existed, neither the stars nor the planets; there was nothing! Then, suddenly, time, space, matter, and energy all came into existence at once in a colossal expansion and the Universe was born.

Light Echo from Star V838 Monocerotis, December 17, 2002. - The variable star V838 Monocerotis underwent an outburst in 2002, during which it temporarily increased in brightness to become 600,000 times more luminous than our Sun. Light from this sudden eruption is illuminating the interstellar dust surrounding the star, producing the most spectacular and gorgeous "light echo" in the history of astronomy. Credit: NASA, ESA and H.E. Bond (STScI).

"Out of a single bursting atom came all the suns and planets of our Universe!" That is a stunningly simple but intriguing idea proposed by a Catholic priest in 1927.

The name of priest, who was also astronomer and professor of physics, was Georges Henri Joseph Éduard Lemaître (1894–1966). He suggested that the Universe started as a small, hot, and superdense "cosmic egg" that expanded very rapidly to the size it now has. Father Lemaître's bold suggestion captured the attention of scientists and astronomers because it explained many mystifying facts that were apparent from scientific observation.

Yet, not everybody liked Lemaître's hypothesis. In fact, many rejected it and others made fun of it. And because Father Lemaître theorized that everything in the world began with an enormous blast, his critics dubbed it the "Big Bang theory." Years later, however, further studies corroborated Lemaître's conjecture and the name *Big Bang* eventually became the accepted term for the scientific theory that explains the origin and evolution of the Universe. Of course we understand that the Big Bang was not a giant explosion, rather it was (and continues to be) a very rapid expansion.

The Big Bang theory describes the birth and expansion of space from an extremely hot and dense state of unknown characteristics, from a "boiling soup" of energy that streamed out, hurling energy in all directions. We can conceptualize this by imagining a balloon expanding, from an infinitesimally small size and growing very quickly to the size of our current Universe.

Within a nanosecond (a billionth of a second), the Universe was hundreds of millions of kilometers in diameter, undergoing a rapid period of cosmic inflation that flattened out nearly all initial irregularities in the energy density. The matter that formed the first stars condensed from this incredibly hot sea of energy. Ever since, the Universe has been expanding and, over time, it became steadily cooler and less dense. Scientists theorize that minor variations in the distribution of mass resulted in the segregation of stars and other matter observed today.

Accepting this theory as truth makes us wonder, when did the Big Bang happen? This question is important because knowing when the cosmic egg of creation burst as described by the Big Bang will tell us the size or the age of our Universe.

Determining the Universe's Birthday

The age of the Universe has been a topic of religious, mythological, and scientific speculation for hundreds of years. In the seventeenth century, Isaac Newton thought that the cosmos was only a few thousand years old. At the turn of the twentieth century, Albert Einstein believed that the natural world was ageless and eternal. Isaac Newton (1642-1727) was the greatest scientist of his era, the genius that gave us the gravitational law and told us how all objects move. Albert Einstein (1879-1955) was the brilliant physicist that changed forever the field of theoretical physics by establishing the laws of relativity, and in the process he altered the way we view the Universe. And yet, neither Einstein nor Newton knew how or when the Universe was born. But let's not be too hard on them, since the answer to the question was difficult to obtain.

For centuries, most people believed that the world was created recently, perhaps a few thousand years ago. In some cultures, the genesis of the cosmos was told from the perspective of religious belief, and many accepted that the world was created in six days in nearly its present condition. Some philosophers speculated the world was perhaps eternal. However, Lemaître's Big Bang hypothesis combined with observational evidence in 1929 seemed to prove all of them wrong.

The realization that we live in a conglomerate of stars we call a *galaxy*, and that there are, in fact, many other galaxies, occurred gradually, thanks to the thinking, observations, measurements, and analysis of many people, including philosophers and astronomers too numerous to name here. However, all the scientific and philosophical speculation that spanned several centuries was finally settled in the first half of the twentieth century, as advances in technology improved the telescopes and other instruments of astronomy—with better tools astronomers could make better observations and see farther.

The first measurements of the spectrum of remote conglomerate of stars were made in 1912. Some astronomers thought that the faint groups of stars were nebulae like the Orion Nebula, but others believed that they were galaxies like ours. At that time scientists didn't know the Milky Way was a spiral galaxy, or that there were many more like it and thousands of conglomerates of stars different and farther away from our Galaxy.

In 1920, a Great Debate ensued between two astronomers, Harlow Shapley and Heber Curtis, concerning the nature of the Milky Way, the Andromeda (known as a "spiral nebula" at the time), and the dimensions of the Universe. Curtis argued that the Andromeda Nebula was actually an external galaxy, and to support his claim, he showed the appearance of dark lanes resembling the dust clouds in the Milky Way, and he noted the significant Doppler shift.

The debate was settled by American astronomer Edwin Hubble (1889-1953). Using a better telescope, Hubble resolved stars that were far too distant, and thus he concluded they were not part of our Galaxy. Eventually, other galaxies were detected. Hubble thus provided the first evidence that there were other "island universes" beyond ours. However, the most perplexing finding was that those galaxies were receding, that is, they were moving away from us!

In 1929, Edwin Hubble and his assistant Milton L. Humason (1891-1972) established that the velocity of recession of a galaxy is directly proportional to its distance. This statement is written with the simple formula $V = H_o \cdot R$., where V is the velocity of recession of the galaxies, typically measured in km/s; R is the distance in mega-parsec (Mpc), and H_o is the constant of proportionality, now known as Hubble's constant.

How did Hubble know the galaxies were moving away? He used the Doppler Effect, the phenomenon that results from the change in frequency of a wave for an observer moving relative to the source of the wave. You are probably familiar with the Doppler Effect which occurs when you hear the sound emitted by a moving source. For example, by the pitch of an ambulance's siren you can tell whether the ambulance is moving towards or away from you. If the ambulance is moving towards you, the sound waves are squeezed together and produce shorter wavelength and higher frequency waves. If the ambulance is moving away, the opposite is true. This phenomenon is known as Doppler Effect, and because light also behaves as a wave, the same occurs with the light coming from a galaxy.

Thus, light from a galaxy that is moving away from us is stretched out. This Doppler shift to longer wavelengths is called a redshift. If a galaxy were moving towards us the light waves would be squeezed, producing a blueshift towards shorter wavelengths. In general, to measure the red or blueshift, astronomers look at the spectrum for known spectral lines that are shifted towards longer or shorter wavelengths compared to what they see in a laboratory. Hubble thus formulated the empirical redshift distance law of galaxies.

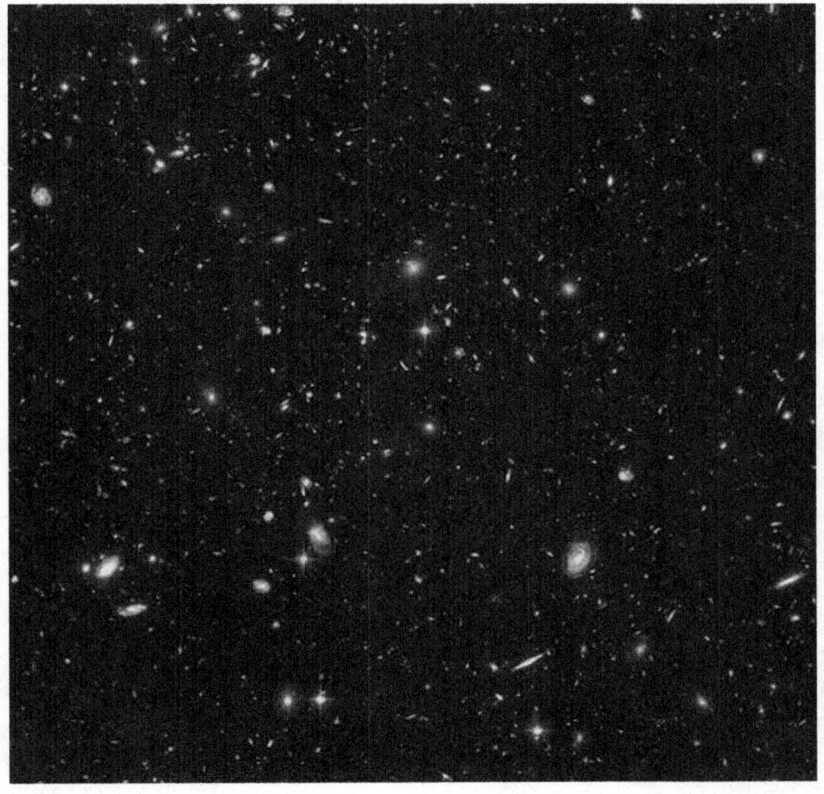

Hubble Ultra Deep Field (HUDF) image obtained by pointing the telescope at one part of the sky for about 11 days. It shows about 10,000 galaxies, some of which may be from the earliest moments when the Universe was born. Credit: NASA, ESA, S. Beckwith (STScI) and the HUDF Team.

This principle, now known as Hubble's law, states that the redshift in light coming from distant galaxies is proportional to their distance. A redshift is a shift in the frequency of a photon toward lower energy, or longer wavelength. Hubble's law is considered the first observational basis for the expanding space paradigm, the evidence needed to support the Big Bang theory. By measuring the redshifts of galaxies Edwin Hubble was able to prove that the Universe is indeed expanding.

In December of 1995, the Hubble telescope, after pointing in one particular part of the sky for 11 days, photographed the deep field or Hubble Ultra Deep Field, HUDF, a region of the size of a thirty millionth part of the area of the sky that contains like ten thousand galaxies, some of which could be from the first moments when the Universe was being born. A similar image of the south hemisphere was taken in 1998, showing notable similarities on both, reinforcing the principle that postulates that the structure of the Universe is independent of the direction in which it is seen.

All astronomical measurements to date indicate that there are no stars older than 14 billion years old. No newly formed galaxies have been detected, implying that all galaxies formed at the same time. Researchers estimated that the oldest star clusters are about 12 to 14 billion years old. So, we can assume that our Universe is at least 14 billion years old. The age is given in years because we are assuming that the light from the farthest visible objects has traveled that many billions of years at the speed of light (299,792,458 m/s), thus giving us the units of years instead of light-years.

Again, what is the age of our Universe? The correct answer is: *it depends.* In the theoretical framework of the Big Bang, the age depends on the Hubble constant and the total mass and energy in it. This is a constant of proportionality found in Hubble's law, which establishes how fast galaxies are moving.

The first calculations made by Hubble were based on the data of redshift of just 46 galaxies, resulting in a value of 500 km/s/Mpc (1 Mpc = 3.26 Mly). Using this value, the cosmos would be just two thousand million years old; that was an erroneous result, since using the isotopes of the rocks determined the Earth was about 4,500 million years old. However, at the beginning of the 1970s the value estimated for H_0 varied from 50 km/s/Mpc to 100 km/s/Mpc, depending on the method used to calculate the Hubble constant. According to these data, the estimated age of the Universe was from ten thousand million years to twenty thousand million years. Clearly, astronomers had to correct such discrepancy.

The errors in the estimate of H_0 were due mainly to the limitations of the instruments. During the second half of the twentieth century, one of the main objectives of cosmology was to refine the value of the Hubble constant.

The Big Bang theory predicts that, at the beginning, the Universe was very hot and that, as space expanded, the gas within cooled down. In 1948, George Gamow and Ralph Alpher speculated that now the sky should be filled with radiation from the remnant heat left over from the Big Bang. The scientific community needed confirmation of that, but how? The Universe itself provided the evidence.

The cosmic microwave background (CMB) radiation was discovered quite by accident in 1965 by American scientists Arno Penzias and Robert Woodrow Wilson. Working for Bell Laboratories, Penzias and Wilson were using a radio telescope for an entirely different purpose when they heard a strange sound that defied explanation; it was as if the sky was whispering. At first the researchers assumed the noisy signal was interference. Later they thought the signal was marred by heat from the bird droppings on the antenna, but after cleaning the antenna, the noise remained. The Universe continued whispering. Eventually, Penzias and Wilson understood the "whisper" and concluded it was indeed the microwave background radiation of the cosmos.

The left over radiation is the best evidence for the Big Bang model of creation and provides data about the Universe primogenitor, including the value of H_o, and by studying it cosmologists have a second method (in addition to the redshift of galaxies) to calculate the Hubble constant.

In 1989, NASA launched the Cosmic Background Explorer (COBE), a satellite designed to observe, map, and measure any lack of homogeneity in the background radiation of space. COBE first determined that, as predicted by the Big Bang theory, the CMB is very smooth and has a black body temperature of 2.725 K. Its overall view of the sky showed that one-half of the sky is slightly warmer than the other half, which was attributed to the motion of the Solar System through space.

In 2001, a NASA exploration team launched WMAP, a space satellite with the purpose of refining the measurements of the background radiation. WMAP, short for Wilkinson Microwave Anisotropy Probe, produced the first full-sky map of the ancient microwaves that fill the Universe. In 2003, researchers published the first results of WMAP, giving a value of 71 (km/s)/Mpc for H_o. In 2006, more detailed analysis resulted in $H_o = 70$ (km/s)/Mpc.

However, observations of far away supernovae revealed that there is another factor that propels the expansion of the cosmos. Astronomers concluded that there is an unknown type of energy, dubbed "dark energy," that accelerates the expansion, and so the age of the Universe, taking into account this acceleration, is closer to 14,000 million years, which is consistent with the age of the oldest stars. This estimate was refined in 2010, and now we know the birthday of our Universe— it is 13.75 billion years old!

How old is the Sun?

Now you may wonder, what is the age of the Sun? Knowing this will also help us determine how old is our planet and, consequently, it will put in perspective the origin of our species.

Our star was born billions of years after the Universe came to be. By observing the processes that occur in the Milky Way and in other galaxies has allowed scientists to formulate one hypothesis about the formation of our Solar System. Most studies suggest that the Sun and all its planets were formed about 4.5 billion years ago.

According to this, the Solar System was formed from the condensation in a nebula by the action of gravity. In the course of this process, matter was organized in the shape of a flat disk with most of the mass concentrated in the center, leading to the birth of a star—what would become the Sun. Smaller mass accumulation formed the planets, moons and asteroids that move in their own elliptical orbit around the central axis. The Earth began forming about 4.6 billion years ago and was completed within 10-20 million years. And so, the age of our planet is about one third of the age of the Universe! Interestingly, multi-cellular life on Earth began a billion years ago, and anatomically modern human beings appeared about 200,000 years ago.

The Sun is just one of thousands of millions stars in the Milky Way (maybe one among 10^{11} stars). And like all stars, the Sun is a massive, luminous ball of plasma, a body of sizzling gases that radiates energy derived from thermonuclear reactions in its interior or core. About 75% of the Sun's mass is hydrogen, while just 2% are elements that we find on the Earth; the remaining mass is helium.

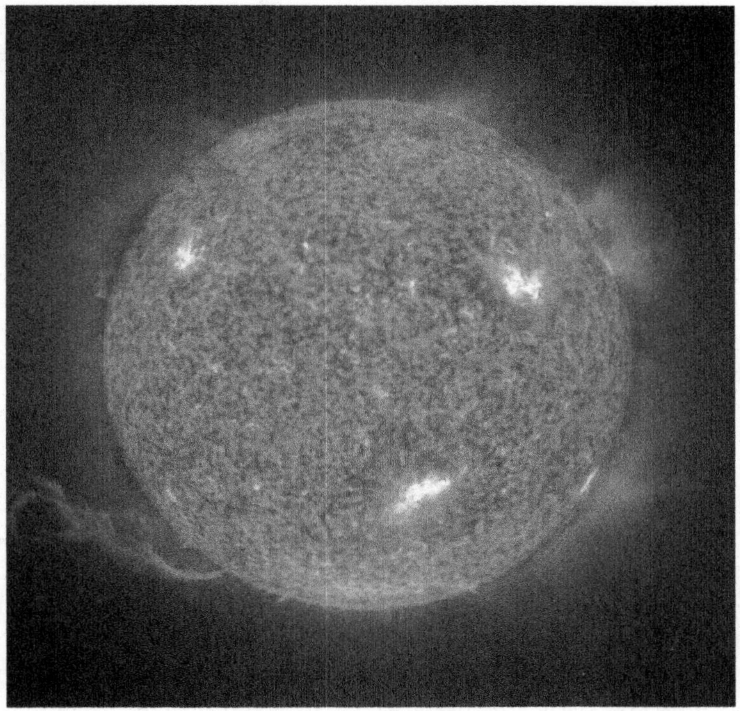

Solar Corona.Credit: Solar X-ray image from the Yohkoh mission of ISAS, Japan. The X-ray telescope was prepared by the Lockheed-Martin Solar and Astrophysics Laboratory, the National Astronomical Observatory of Japan, and the University of Tokyo with the support of NASA and ISAS.

The core of the Sun is a searing hot furnace at very high temperatures due to the thermonuclear processes that occur in its interior. The Sun's energy is generated in the deepest part of its core by means of one of the most powerful process in the Universe: nuclear fusion. The nucleuses of hydrogen are torn, forming helium and liberating enormous quantities of energy. That is the reason a star is so bright. It burns its fuel by nuclear fusion (different from the fire, which burns due to oxidation). The equilibrium between the pressure outwards of gas heated by fusion and the attraction inwards of gravity is known as hydrostatic equilibrium.

The core of our star is surrounded by radiative layers and convection layers. In the radiative zone, closest to the core, the gas is quiet and steady, and its energy (light in all wavelengths) is diffused through the layer as radiation. Over this layer there is the convection where turbulent flow of gas carry the solar energy outwards in a process called convection: the gas is simultaneously heated from below by fusion and it is cooled from above since the energy is liberated and dispersed through space. Convection makes the gas churn and rotate in agitated motion, just as water before it boils.

The Sun has a photosphere, a layer with an average temperature of 6000 K and from where most of the energy that arrives to Earth is generated. A solar corona envelops all layers of the spherical Sun. Made of ionized gases at temperatures near a million degrees, the corona extends to millions of kilometers by the solar wind. The corona is peculiar; it is much hotter (by a factor of nearly 200) than the visible surface of the Sun. It makes us wonder, *Why is the Sun's Corona so much hotter than the Sun's surface?* Nobody knows, although there are several theories to explain it. This is now one of the unsolved problems in physics.

What we know is that almost all elements heavier than hydrogen and helium are created inside the cores of stars. And thus, we (you and I and everybody else) are made of stars. Because we are made of atoms and those atoms come from the stars! Gold, too. The precious metal we like so much also comes from stars.

All gold found on Earth started out in the center of a star—a supernova to be precise. A supernova is an exploding star, the dying event that occurs when the cores of massive stars—more than seven times as heavy as the Sun—collapse under their own weight and trigger an explosion. After the explosion, gold atoms are cast deep into space where they stay for eons. Eventually, some gold atoms may be part of a cloud that will become a planet. Once inside the planet, the gold near the surface can be extracted, as people on Earth have been doing for centuries. So, if you have a gold ring or wear gold earrings, you can thank the stars for your good fortune. We can thank the heavens for our existence!

We owe our life to the Sun. Life as we know it would not be possible without the heat and light of this marvelous cosmic body that controls our Solar System. Even more intriguing is the fact that our existence is possible due in part to Earth's distance from the Sun, allowing water to exist in all three states, and in a cycle. Water is essential for life. If we were closer to the Sun but not as close as Venus, for example, life may still have evolved, but perhaps differently. On the other hand, if we were a little further away, say at the distance of the gaseous planets, who knows? The fact is that we owe our existence to the Earth being at the right distance from our mother star: not too far, not too close.

What would happen to Earth, and to the human race, when the Sun stops shinning? Like all other stars, the Sun will die one day, in about 5 billion years from now. That is predicted because its age today is about 4.57 billion years and during that time it has consumed almost half the amount of hydrogen in its core (hydrogen is the fuel that drives nuclear fusion inside).

Just like humans, all stars are born, live for millions of years, and finally stars die, ending up as white dwarfs (stellar remnant), neutron stars, or as black holes. Massive stars end their lives in spectacular cosmic explosions that we call supernovae. The Sun being a small star does not have enough mass to explode as a supernova but it will become a red giant when all its fuel is consumed. Gradually, it will expand, reaching a maximum radius beyond the Earth's current orbit (1 AU) before dying and becoming a white dwarf.

What will happen to Earth when the Sun becomes a red giant star? Will it be swallowed by the layers of solar gases? Most probably; unless people find a way to move the Earth to a farther orbit. If not, the Sun will literally gobble up our world, as other suns have been observed eating their planets that gravitate too close. That occurs because the gravitational force of the mother star can trap the gases from the planet's atmosphere and gradually eat the entire world. But that's another problem to be solved in the future.

The sky is full of stars, some are relatively young, others are about to be born, and some may be about to end their lives. The most massive stars have the shortest lives. Stars that are 25 to 50 times that of the Sun live for only a few million years. They die so quickly because they burn massive amounts of nuclear fuel. And when these huge massive stars die, they become supernovae. A supernova is a massive star that explodes and becomes extremely luminous.

Every fifty years or so, a luminous burst of radiation in a typical galaxy like the Milky Way, announces the death of a massive star. Supernovae are very important for the evolution of the Universe. They play a significant role in enriching the interstellar medium with higher mass elements. Furthermore, the expanding shock waves from supernova explosions can trigger the formation of new stars. Also, astronomers realize that the maximum intensities of these explosions can be used as distance indicators.

A giant Hubble Mosaic of the Crab Nebula. The Crab Nebula, Messier 1 (M1, NGC 1952), is the most famous supernova remnant, the expanding cloud of gas created in the explosion of a star as supernova that was observed in the year 1054 AD. It shines as a nebula near the southern "horn" of Taurus, the Bull. Credit: NASA, ESA, J. Hester and A. Loll (Arizona State University).

The explosion of dying massive stars is a relatively rare phenomenon. It is estimated that each galaxy produces, on average, a supernova every six centuries. The oldest record of a supernova was observed in China in 185 B.C. One of the most famous was recorded by Chinese astronomers in 1054 A.D. in the constellation of Cancer. Today, the shattered remains of the star appear as a cloud of gases, known as the Crab Nebula. The supernova of 1054 was identified by Edwin Hubble in 1928. Using the expansion velocity of Messier 1, the Crab Nebula, Hubble deduced that the nebula was born 900 years ago. This approximate birth date correlated with the historical record of the appearance of a new or "guest star" reported by the Chinese.

In 1604, before the telescope, astronomer mathematician Johannes Kepler (1571-1630) discovered a "new star" that was "brighter than anything else in night sky except Venus." What Kepler saw was actually the end of a massive star as it exploded and became extremely luminous in the process. But Kepler did not know that, and because the brilliant point in the sky was not there before Kepler called it *stella nova,* which means, literally, new star.

Kepler documented the new point of light near the foot of the constellation Ophiuchus and kept very detailed records until the star faded from view. He published his observations in a book with the title *De Stella Nova,* drawing an image of the constellation to show the position of the new star (noted with an "N" on the foot of Ophiuchus). Kepler's supernova was the second to be observed in a generation (after one recorded by Tycho Brahe in Cassiopeia). As far as I know, no supernovae have since been observed with certainty in the Milky Way. However, with the advent of modern, more powerful telescopes in the twentieth century, supernovae in other galaxies have been observed.

The remnant of Kepler's supernova (Supernova 1604) was found over three hundred years after the star's explosion. A search for the remnant of Nova Ophiuchi led to its discovery by astronomer W. Bade, who in 1943 reported it as "a small patch of emission nebulosity, which is undoubtedly a part of the masses ejected during the outburst." Between June 2000 and August 2004, NASA's three Great Observatories—the Hubble Space Telescope, the Spitzer Space Telescope, and the Chandra X-ray Observatory—were used to prove the expanding remains of Kepler's supernova.

Supernova remnants greatly impact the ecology of the Milky Way. If it were not for supernova remnants, we would not be here, as there would be no Sun or Earth.

Pondering about the birth and future of our Universe

According to supporters of the Big Bang theory, our Universe was born out of a "singularity" around 13.7 billion years ago. What is a "singularity," you may ask, and what caused it? Well, scientists don't know for sure. Physically, a singularity is a point in spacetime which defies understanding. A singularity is thought to exist at the core of black holes. We will explain further in chapter 7; for now we will only say that black holes are massive cosmic objects of intense gravitational pressure. The pressure is so powerful that matter is actually compressed into infinite density in the center of the black hole. These points in spacetime of infinite density are thus singularities. So, according to the Big Bang theory, our Universe began as an infinitesimally small, infinitely hot, infinitely dense cosmic egg, also called a singularity.

It seems to me that the Big Bang theory can only explain what happened a moment after the cosmic egg burst into being and its subsequent expansion. But it cannot answer the questions, where did the egg come from? We don't know. Why did it appear? We don't know either. The theory only tells us the Universe had a beginning. According to the Big Bang theory, our Universe was born billions of years ago and has been expanding ever since. We also don't know whether it will continue growing and eventually break down and die. Maybe it will never die, but for sure we will never know. But we can theorize and speculate how it may end. It all depends on what kind of model of the Universe we accept.

There are three cosmological models of our Universe. The first is that the universe is "closed," the second is that it is "open," and the third is that it is "flat." If it is closed, and one would try to reach the edge of the universe, one would eventually loop back to where one started. It would be like walking on a closed circuit or round track—eventually one return to the place where we started.

If the Universe were open or flat, one would never reach the outer edge, as space would continue on infinitely. The difference between open and flat is that an open universe would be curved. In either case, if two beams of light were traveling into space, starting initially parallel to each other, in a flat universe both light beams would remain parallel, while in an open universe they would actually get farther from each other.

If the Universe were closed, one would imagine that it could be inside another universe, like a matryoshka doll or Russian nesting doll (a set of dolls of decreasing sizes placed one inside the other), as some scientists theorize. But then again, how was the other universe born? This is like a never-ending story, exciting, but eventually we get tired and wish to reach the end, a conclusion of sorts. And we continue asking, how that initial bursting atom originated? Where and how the cosmic egg was hatched?

A new theory postulates that the natural world may have started not with a Big Bang but with a "big bounce." What some contemporary scientists are saying is that the Universe started with an implosion that triggered an explosion or expansion driven by quantum gravitational effects. This idea originated from the conclusion that the Big Bang singularity—an infinite density—is unrealistic. Maybe they are correct. But since the "big bounce" theory, as intriguing as it is, has not been verified, we leave it for another discussion.

Yet another theory says that our Universe might have originated from a black hole that lies within another universe—assuming the closed universe model. This concept arouse from the idea that matter and energy falling into a black hole could (in theory) come out as a "white hole" in another universe. In such a situation, both the black hole and the white hole would be the mouths of an Einstein-Rosen bridge, also known as a wormhole. We will touch on those ideas later.

Nevertheless, even accepting the Big Bang theory, the age of the Universe remains an open topic of study by astrophysicists and cosmologists. Both astronomical observations and experiments on particle super colliders underground have the goal of providing evidence to support the main premises.

The particle supercollider at CERN, the European Organization for Nuclear Research near Geneva (Switzerland), is preparing for the much anticipated experiments that will recreate the rapidly changing conditions in the Universe a split second after the Big Bang. It will be the closest that scientists can come to simulating the birth of the Universe. The new 27-km (17 miles) long supercollider, known as Large Hadron Collider (LHC), was built in a tunnel 100 m

(300 feet deep) underground. Scientists hope the new LHC will enable them to study particles and forces as yet unobserved.

Once the LHC is in full operation, two streams of protons will be whipped up in opposite directions around the underground ring of super-cooled pipes to a whopping 99.999999 percent of the speed of light. When the two waves of protons collide with each other, particles will transform into bits of energy up to 100,000 times hotter than the Sun's core—a process that would replicate what the Universe was like just an instant after it was born. CERN will become, literally, the center of the universe.

3
A Cosmic Paradox

Why is the night sky dark?

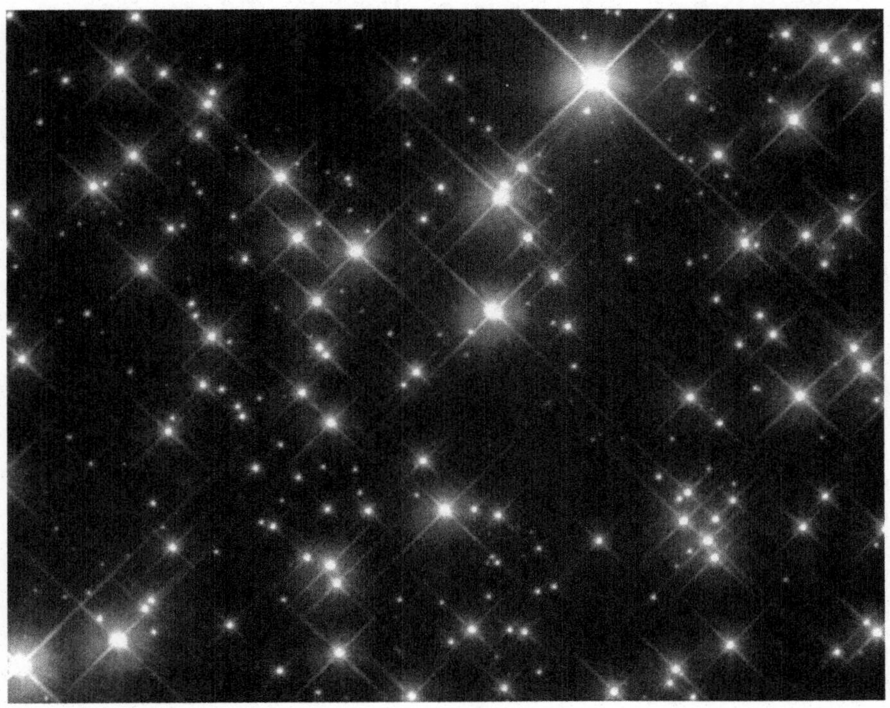

Ancient white dwarf stars in the Milky Way Galaxy. Credit: NASA and H. Richer (Univ. of British Columbia).

Really! If there are trillions of stars, don't you wonder *why the sky is so dark at night?* Should not those many stars keep the sky aglow with light? The sky should be bright everywhere, not black with tiny dots of light. This paradox baffled the most brilliant scientists for centuries. Kepler, Newton, and Einstein were among the great scholars who could not answer such child-like question.

Johannes Kepler was the German astronomer who gave us a beautiful and powerfully predictive description of the Solar System: the laws of planetary motion now known as Kepler's laws. Isaac Newton derived the universal theory of gravity and gave us the physics that explained Kepler's laws and showed that they were not perfect. And Albert Einstein was the theoretical physicist who, with his law of general relativity, gave us a better explanation of what gravity is, showing that Newton's concept of gravity wasn't quite right either.

Newton had a sense of an infinite universe, a necessary idea for him to explain how the gravitational forces acting between stars did not pull them all together into a giant blob of mass and light. But if space stretches on forever, as many people thought, every line of sight must eventually intersect a star, right? If this were the case, stars would appear to overlap each other obliterating all darkness.

In the 19th century, the German amateur astronomer Heinrich Olbers raised once again the question of the dark night sky; he thought something was wrong with the idea of an infinite and eternal static universe—the accepted concept of his time. Olbers (1758-1840) was a physician but he is better known for his contribution to astronomy. He converted the upper floor of his home into an observatory and spent his free time there, watching the night sky. From that work Olbers devised the first satisfactory method of calculating the orbits of comets. In 1802, he discovered asteroid Pallas, and on March 29, 1807 he discovered asteroid Vesta (named after the Roman virgin goddess of home and hearth, Vesta). As the word "asteroid" was not yet coined, the literature of the time referred to these newly discovered bodies as planets.

Olbers' observation about the darkness of the night sky, now known as Olbers' paradox says that in an infinite Universe filled with stars, the sky should be a continuous blaze of light, yet the night sky remains dark. Why? It seems that one acceptable answer was given by Edgar Allan Poe, who wrote in one of his essays of 1848: "light from faraway stars has not reached us, and that is why the night sky is dark." It's pretty amazing conclusion for a writer at that time, don't you think? It took a long time for scientists to provide another answer.

In the twentieth century, Einstein proposed that space and time were dynamic and changing, not the static background believed at the time. He shocked the world saying that space is not flat and gave us instead the notion of a curved spacetime. Einstein's general theory of relativity changed our understanding of gravity, introducing the radical idea that the Sun (or any other cosmic object) bends spacetime, and spacetime tells planets and all other objects in the Universe how to move.

As a consequence of gravity distortion, Einstein concluded that the light from stars would "bend" as it passes by the Sun, thus shifting the relative position of the stars ever so slightly. This phenomenon is known as gravitational lensing or bending of light. Not every scientist embraced this radical idea, at least not until it was irrefutably proved. Einstein proposed a test in the largest laboratory we have: the sky. The 1919 solar eclipse provided the opportunity for a British expedition led by Arthur Edington to test Einstein's prediction.

Edington compared photographs of stars from the Hyades star cluster seen in the vicinity of the eclipsed Sun, with photographs of the same stars when the Sun was off the visual field. The photographs confirmed the predicted shift in the stars' apparent position. This amazing result proved that indeed gravity distorts space, and it turned Einstein into a superstar, perhaps the first scientist to become a celebrity. He won the Nobel Prize in Physics in 1921, although not for that achievement nor for his revolutionary relativity theories.

In the midst of an exciting period of planetary exploration in the 1970s, space probes also helped to verify that space *is* curved. When two *Viking* spacecraft where orbiting Mars, scientists observed that, as Mars hid behind the Sun, the robots radio signals were suddenly delayed by several microseconds, as if the Red Planet and the space probe had abruptly jumped 18 miles farther away from the Earth. The scientists concluded that it was caused by the longer path taken by the signals when they traversed the distorted space.

But much before that, Einstein developed a set of powerful equations that describe exactly how spacetime bends and quivers. The so-called Einstein's tensor, describing the curvature of spacetime, is a deceivably simple mathematical relation, as it unfolds into several equations. With these equations, Einstein predicted that the Universe either expands or contracts. Believing that the solution that gave the option for space expansion was not acceptable, Einstein made a mistake by adding a term to his equations that forced the solutions to yield a Universe constant in size. Years later, Hubble discovered that the dozen or so known galaxies were running away from each other, receding from the Earth at great speeds—some up to 2 million miles per hour (3.2×10^6 km/h), suggesting that the Universe is in fact expanding. Einstein's blunder was corrected with this fascinating discovery.

Now Olbers' paradox is explained on the basis of the age and evolution of the Universe; that is, from the combined observation that (1) the speed of light is finite, and (2) the Universe has a finite age, i.e. we only see the light from parts of the Universe less than 15 billion light-years away. Put another way, since we know that galaxies are distributed sparsely enough in the cosmic vacuum, in most directions there are no stars, and thus the sky is dark. At some enormous distance away, some galaxies are receding and cannot be seen. And since the Universe is not infinite in time or space, nothing can be seen beyond 15,000 million light-years—the size of the Universe is just 13,750 million light-years.

We measure age in units of time. And so you may ask, what is *time*? The concept of time itself is difficult to define. Did time originate with the Big Bang? What if there was time before the Big Bang? If the initial expansion was from a pre-existing physical state, one can conclude that there was never a beginning. In such scenario, the Big Bang could be just one cosmic explosion among others throughout an infinite past. But then we have to deal with the concept of infinity.

For now, let's review some tested scientific laws that may help us understand somehow the nature of time.

4
A Question of Time

Time is a fourth dimension, tightly linking reality with space

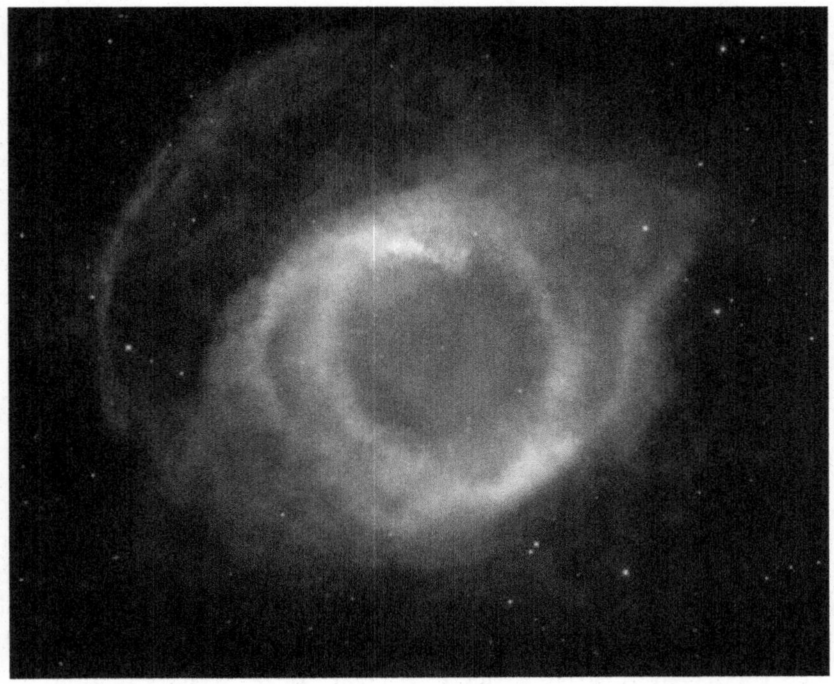

The Helix Nebula: A gaseous envelope expelled by a dying star. Credit: NASA, ESA, C.R. O'Dell (Vanderbilt University), M. Meixner and P. McCullough (STScI).

We assume time is unidirectional and with a fixed origin at the instant of creation. Is this correct? If time is intrinsically linked to the birth of space we say that "Time is a created dimension." But the question of the nature of time cannot be answered lightly. Time is a way of separating events from each other. We perceive time when we see things change, when we age, and when we observe nature changes moment by moment. Change is an intrinsic feature of the physical world, so we also say that time is what's measured by clocks. That does not answer the fundamental question either.

The notion of time has been studied by philosophers, theologians, and scientists for thousands of years. Yet, many issues remain unresolved. The Big Bang theory implies that time began with the birth of Universe. At birth, time, space, matter, and energy all came into existence at once. Before that, before the Big Bang, time did not exist! Accepting this as a fact we're startled by the realization that our Universe could not exist without time, and time could not exist without the Universe; they both are different components of one entity we call *spacetime*.

In physics, time is measured by motion and it becomes evident through motion. In the sixteenth century, Isaac Newton defined mathematically how motion changes with time. In his book published in 1687, *Philosophiæ Naturalis Principia Mathematica*, better known as the *Principia,* Newton showed that the force causing apples to fall is the same force that drives planetary motions and produces tides. However, Newton could only describe how gravity operates instantaneously at a distance. Space was a fixed, infinite, an unmoving metric against which absolute motions could be measured. He assumed the cosmos was ruled by a single absolute time represented by an imaginary clock.

After Newton formulated the law of universal gravitation and the laws of motion, scientists thought that the positions and motions of bodies in space should all be measurable relative to some non-moving, absolute "frame of reference," which was assumed to be filled with an invisible substance called "the ether." In 1905, Albert Einstein rejected the Newtonian view of space and time and replaced it with the special theory of relativity, called "special" because it is restricted to frames of reference in constant, unchanging motion—because they are not being accelerated by a force.

Einstein based his special theory of relativity on two principles. The first, called the principle of relativity, states that the same laws of physics apply equally in all constantly moving frames of reference. The second principle states that the speed of light is constant and independent of the motion of the observer or source of light. One of the fundamental implications of special relativity is that space and time are intimately linked!

While spacetime is viewed today as a consequence of Einstein's special relativity theory, it was first explicitly proposed mathematically by one of Einstein's professors, Lithuanian-born mathematician Herman Minkowski (1864-1909). Between 1907 and 1908, Minkowski formulated a view of the world in which the traditional three dimensions of space were combined with the additional dimension of time. This concept was written explicitly in his lecture notes. And so it seems that Minkowski was the first to realize that Einstein's ideas of special relativity could be best understood as a coupled four-dimensional *spacetime continuum.*

What Minkowski meant is that space and time do not exist independently of one another. We and every object everywhere in the Universe travel through a combined space-time plane. Even if we are standing, the Earth is moving, and so we are continuously moving in time. We grow old and sense time by the passing of days, and we cannot stop that change.

The spacetime continuum provided a framework for all later mathematical work in relativity. Einstein then used Minkowski's ideas to examine the connection between gravity and acceleration. In 1915, Einstein published his theory of general relativity, which provided a new description of gravity. With this theory, and with his set of field equations, he described how mass distorts spacetime—the concept of gravitational lens explained earlier.

Einstein's most famous equation, $E = mc^2$, expresses the equivalence of energy (E) and mass (m) via the square of the speed of light (c). With this formula Einstein suggested that mass and energy are distinct forms of the same thing. He included it in his essay, "Does the inertia of a body depend upon its energy-content?" published in *Annalen der Physik* on September 27, 1905. Between 1907 and 1915, Einstein developed the general theory of relativity, and a theory of gravity more accurate than Newton's.

In February 1917, Einstein published the paper "Cosmological Considerations on the General Theory of Relativity," in which he reflected on the effects of matter and energy on the actual geometry of the Universe. In 1922, Russian mathematician Aleksandr Friedmann (1888–1925) published a set of solutions for Einstein's general relativity equations. Friedmann concluded that there was no unique solution to Einstein's equations, rather there was a whole family of possible solutions. This family of solutions thus allowed for different cosmological models of the world.

In Friedmann's models, the only force that is considered is gravitation, resulting in universes that are homogeneous (the same everywhere on a large enough scale) and isotropic (that looks the same in every direction). Most importantly, these models incorporate the idea of space expansion and, in some cases, contraction. Einstein himself had viewed the Universe as static. Friedmann thus provided the theoretical framework for an expanding Universe within the spacetime and mathematics of general relativity.

A few years later, Father Lemaître derived the same solutions, unaware of Friedmann's earlier work. This was around the same time when astronomers discovered many more conglomerates of stars far away from the Milky Way. Moreover, Hubble discovered that the galaxies were receding. Lemaître realized that the newly discovered galaxies could be used to prove the expansion of the Universe and proposed that, by going back in time, one would arrive at the moment of creation.

Through the remaining decades of the twentieth century, astronomers observed that distant galaxies appear different; they are bluer and more spiral-shaped compared to the conglomerates of stars closer to our own. This suggests that the Universe has indeed evolved since it was born. Today, the cosmos appears to be increasing its rate of expansion, and it's unclear when or whether it will stop growing.

Time Paradox

One of the most fascinating ideas of modern physics is the paradox that deals with the effects of time in the context of space travel at velocities approaching the speed of light. Einstein's theories of relativity state that time cannot be treated absolutely separately from space, only in one observer's relative view. What this means is that time is relative to the speed a body is traveling. Einstein utilized the example of two watches: one at rest and the other moving. He proposed that the watch transported at velocity near the speed of light would move slower as compared with the watch remaining at rest. This phenomenon is known as "time dilatation."

Time dilatation is simply described to mean that an observer sees that another watch (an identical watch to hers) is measuring time at a slower rate than hers. This is normally interpreted as if *time has slowed down* for the other watch, but that is true only in the context of the system of reference of the observer. Locally, time is moving at the same rate. The phenomenon of time dilatation applies to any process that manifests changes through time.

The watches paradox is now described with the analogy of the twin sisters. If a twin would travel in a spaceship moving at a velocity near light speed while the other twin stayed on the Earth, the sister at rest would age faster than her traveling twin, according to the paradox. Known as the "twin paradox," it is a mental experiment used to analyze the different perception of time between two observers with different states of movement. The system of the stationary twin on Earth experiences effects different from those of the system of her twin traveling onboard a moving spaceship. While the twin on Earth sees the vehicle in space moving at velocity v, the twin traveler on the spaceship has a relative speed of 0.

Einstein's theory of special relativity postulates that the measure of time is not absolute, and that, given two observers, the measured time between two events by those observers does not coincide but that the different measures of time depend on the state of relative motion between them. In the theory of relativity the measures of time and space are relative and not absolute because they depend on the frame of motion of the observer. It is in this context that the twin paradox is formulated.

In our everyday life, in the macroscopic world we observe, time is represented with an arrow pointing from past to future. Perhaps because we can see into the past but can only predict the future. In fact, we are able to determine when the Universe was born—within a nanosecond—but we cannot say with all certainty whether it will die and can only predict if it space would stop expanding one day.

Famed British astrophysicist Stephen Hawking announced publicly that he believes humans could travel millions of years into the future. What Hawking is saying is that time travel is possible, but only into the future. Moving backwards is impossible, Hawking states, because it "violates a fundamental rule that cause comes before effect." Also, I say, we need a very powerful time machine, a space vehicle that can travel at higher than light speed.

According to Hawking, if spaceships were built that could fly faster than the speed of light, a day on the ship would be equivalent to a year on Earth. Thus, it would take six years at full power just to reach these speeds. In Hawking's scenario, "after the first two years, the spaceship would reach half light speed and be far outside the Solar System. After another two years, it would be traveling at 90 per cent of the speed of light. After another two years of full thrust, the spaceship would reach full speed, 98 per cent of the speed of light, and each day on board would be a year on Earth. At such speeds, a trip to the edge of the galaxy would take just 80 years for the astronauts traveling on the superfast space vehicle."

Hawking's theory may be tested in experiments at a large particle collider such as the Large Hadron Collider in Geneva. According to a scientist, when they accelerate tiny particles to 99.99 per cent of the speed of light, the time they experience passes at one-seven thousandth of the rate it does for the observers. Of course this is true for fundamental particles. What about an astronaut on a superluminal space vehicle?

Let's for accept for now that Hawking is correct and contemplate a future when humans could travel millions of years into the future and repopulate the Earth, if still intact, or search for another planet to start a new civilization. Because, as we said earlier, the Sun will not live forever and neither will the world we know!

5
Mysteries Unsolved

The cosmos is full of wondrous and magical secrets, waiting patiently for us to unravel

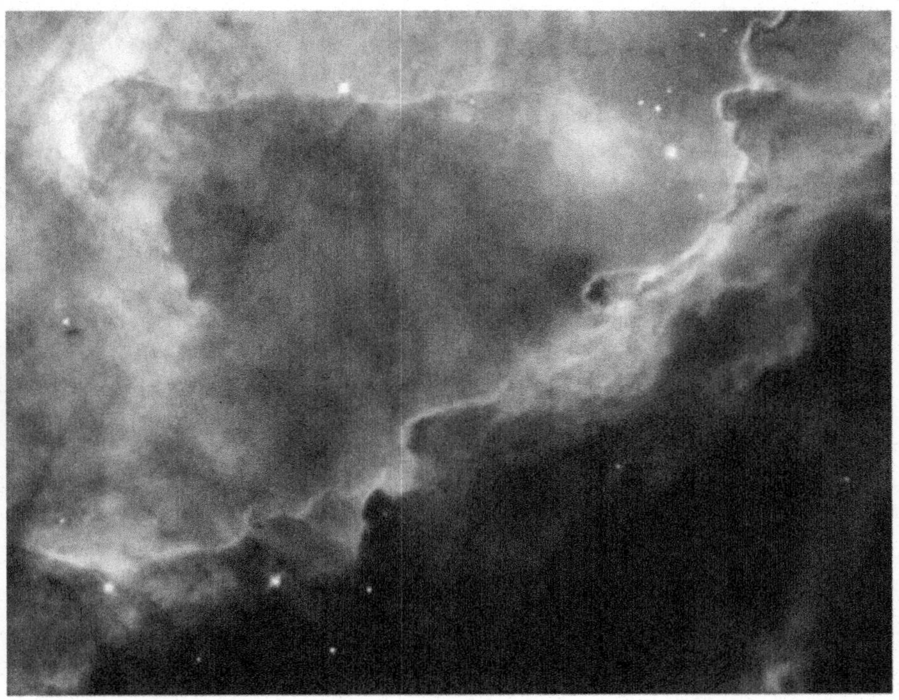

A Perfect Storm of Turbulent Gases in the Omega or Swan Nebula (M17). Also called the Horseshoe Nebula, the Omega Nebula is a region of star formation and shines by excited emission, caused by the higher energy radiation of young stars. Credit: NASA, ESA and J. Hester (ASU).

Our world is made mostly of material we can't see and is fueled by energy we can't measure. Sure, we know the Universe contains energy and matter in many forms: dust and gas, the cosmic clouds where new stars are forming, moons and comets, planets and stars. But there are many more things hidden from the scrutiny of our eyes or from the proving of our telescopes and space robots.

There is so much invisible material that scientists call it "dark matter," and the powerful energy that drives the acceleration of the cosmos is known as "dark energy." In fact, about 96 percent of all mass and energy in the Universe is comprised of this dark stuff.

Dark matter is matter that cannot be detected because it does not emit light or any other known form of radiation. Dark matter is only detected through its gravity. In 1998, scientists determined that something is out there, filling up space and adding to the gravity budget of the cosmos. They arrived at this rather radical conclusion because, although it cannot be seen directly, without this as-yet-unseen dark material, they think galaxies wouldn't hold together.

What makes astrophysicists think that this unseen matter exists at all? Well, they infer its presence indirectly from the observed motions of stellar, galactic, and galaxy cluster and superclusters. Dark matter is also needed in order to enable gravity to amplify the small fluctuations in the microwave background enough to form the large-scale structures observed in the cosmos today.

So what do you think is dark matter? Could it be the material that makes up neutron stars, dead stars, black holes? Or maybe dark matter is composed of exotic particles scientists have yet to discover?

Dark energy is another fascinating mystery. The Universe is accelerating at an ever-increasing pace; something invisible is serving in an anti-gravity capacity over large distances, quite literally pulling the cosmos apart. What is this powerful force? This is one of the thrilling secrets waiting to be unlocked. The nature and genesis of dark energy is unknown, but it is believed to be related to vacuum fluctuations generated somehow by space itself. This vast and mysterious dark energy appears to gravitationally repel all matter and causes the Universe to expand.

Dark energy entered our scientific vocabulary in 1998. After making a survey of exploding stars (supernovas) in several distant galaxies, astronomers found that the supernovas were dimmer, bluer and more spiral-shaped than they should have been, and that meant the galaxies were farther away than they should have been. The only way for that to happen, the astronomers realized, was if the expansion had accelerated at some time in the past.

Before that, scientists believed that the cosmic expansion was gradually slowing down, due to the gravitational pull that individual galaxies exert on one another. But the supernova results implied that some mysterious force was acting against the tug of gravity, causing galaxies to fly away from each other at greater speeds. As we said earlier, some galaxies recede at velocities of more than three million kilometers per hour.

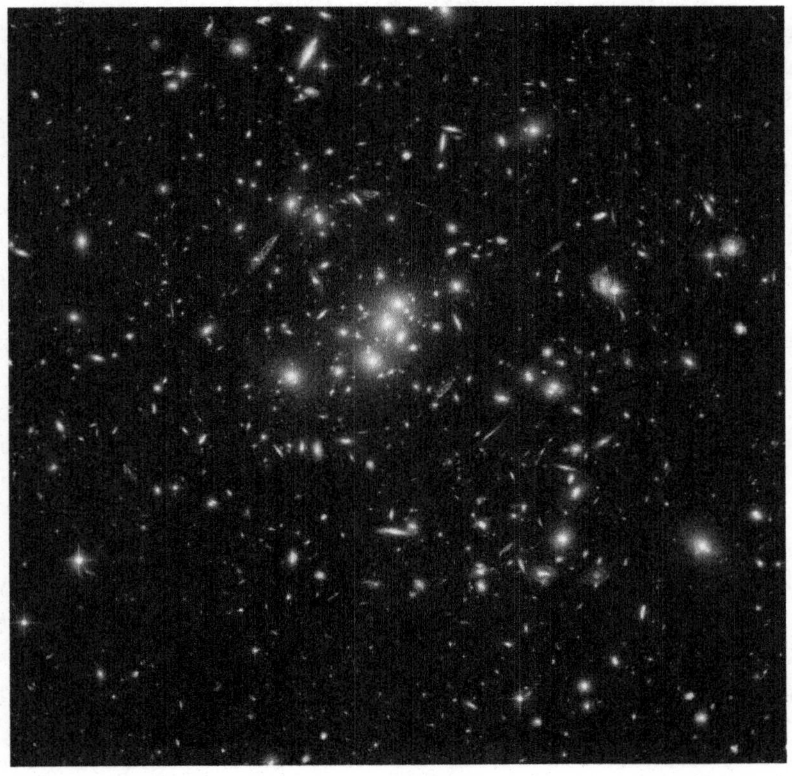

This image shows blue smears representing more distant galaxies, whose light is smeared by the presence of dark matter. Researchers prepared an animation of this image that shows the distribution of dark matter they derived based upon the distortion of galaxies. Credits: NASA, ESA, M. J. Lee, and H. Ford (Johns Hopkins University).

Controversy still exists in the determination of the mass density of the Universe and its corresponding geometry, which scientists call Omega and Lambda, respectively. After interpreting observational results from space-borne and ground-based telescopes, some astronomers began to claim that the Universe has a significant amount of dark energy. Some scientists were intrigued but skeptical. In a now-famous 1998 debate, P.J.E. Pebbles and M.S. Turner defended opposite views. Even today some astronomers doubt that dark energy exists.

The biggest galaxy survey ever made, the Two-Degree Field Galaxy Redshift Survey, at its completion in 2003 produced a map showing the locations of more than 221,000 galaxies. Astronomers used the information in the map to make the most precise estimates of the mass and density of the Universe, and also its relative amounts of normal (baryonic) matter, dark matter, and dark energy.

From that data, scientists predict that the Universe is made up of about 21 to 25 percent dark matter and 70 to 75 percent dark energy. It means that most of the "stuff" in the cosmos is *dark*. Ordinary matter, including stars, planets, cosmic dust and free hydrogen and helium, comprise a measly five percent of the total stuff in the Universe. It is a stunning conclusion!

In 2007, the Hubble space observatory used gravitational lensing to infer the existence of a large distribution of dark matter in a distant galaxy cluster. Gravitational lensing is the displacement of light due to the warping of space by a gravitational lens (a massive object in space such as a galaxy) that bends light traveling across space (as predicted by Einstein). The distortions appeared to occur far away from the actual location of the cluster, indicating—scientists say— that dark matter can extend much farther from the galaxy.

Recently, other researchers proposed that dark matter could actually be shinning with its own kind of light, a dark radiation that we have no technology to detect. This idea seems plausible, especially considering the notion that the diffuse halo surrounding the Milky Way would produce a lot of gamma rays from dark matter annihilation.

Many researchers are working in deep underground laboratories around the world, hoping to detect the components of dark matter. Most experiments are searching for theoretical particles called WIMPS or Weakly Interacting Massive ParticleS—the leading dark matter candidate. Recently, a group of researchers at the Fermilab near Chicago reported that the accelerator produced particles that they are unable to explain and think that it could be connected to dark matter. And there may be other particles as yet undiscovered.

The nature of dark matter and dark energy is probably the most profound mystery in physics, the most difficult of questions scientists must answer. When they do, it will improve our understanding of matter, space, and time. Whoever explains the nature of dark matter would solve one of the greatest puzzles of modern science.

But what if something else accounts for the apparent expansion of the Universe? What if there is no dark energy at all, but rather we—humans on Earth—perceive something that is not so? According to some scientists, Earth may be "trapped in an abnormal bubble of spacetime." What some researchers suggest is that, if we were in a sparse area of space, then cosmic objects could appear to be farther away than they really are. In such case, there would be no need for blaming dark energy as the cause for the acceleration of galaxies, since maybe what we see is the result of some form of optical illusion. The only way to settle this controversial theory is to test it. And scientist will do just that with a proposed "Joint Dark Energy Mission."

The NASA/DOE Joint Dark Energy Mission (JDEM) is a probe that will focus on investigating dark energy. The mission was designed to measure how the expansion rate of the Universe has changed over time by studying Type Ia supernovae: the explosive deaths of white dwarfs. Type Ia supernovae are relatively uniform in their luminosities and other properties, and they are extremely luminous. So they are ideal for measuring distances to remote galaxies.

Another puzzle concerns the distribution of matter. There is scientific debate among researchers, some claiming that the distribution of matter in the Universe is smooth and homogeneous, and others argue that the mass distribution is hierarchically structured and clumpy, like a fractal. Who knows for sure? Research continues.

Quasars, discovered more than 30 years ago, are some of the most mysterious objects in the Universe. It is believed that they are powered by super-massive black holes of several million solar masses or more that are at the centers of remote galaxies. The luminosities of quasars are much higher than ordinary galaxies like the Milky Way, yet originate from regions smaller than the size of the Solar System. Occasionally, quasars eject giant jets of gas that appear to move on the plane of the sky with velocities exceeding that of light (that is, with superluminal velocities). The enormous distance of quasars is intriguing to scientist trying to interpret the source of energy and the nature of the energetic gases that seem to be moving with superluminal speeds.

Another intriguing phenomenon that captivates our attention is the cosmic explosions, formally known as "gamma-ray bursts." A gamma ray burst (GRB) is a short blast of gamma-rays from a given location in the sky whose origin is still mysterious. Scientists think that supermassive stars may be ticking bombs ready to produce gamma-ray bursts of titanic power that could annihilate the Earth. Other researchers believe that every gamma-ray burst may herald the birth of a black hole.

Gamma-ray bursts seem to be more energetic than a supernova, and they are totally unpredictable. The explosions, some brief and some longer, appear from any and all parts of the sky at random and unexpected times. In 2004, NASA spacecraft Swift detected hundreds of bursts, monitored their glowing debris at multiple wavelengths, and measured their distances. Astronomers determined that the blasts were 12.8 billion light-years away—that means the explosions originate near the outer edge of the observable Universe! The sources of these cataclysmic high-energy flashes remain a mystery.

In 2008, NASA scientists announced that they had detected a new kind of pulsar, one that "blinks" in pure gamma-rays. The 10,000-year-old stellar corpse is sweeping a beam of gamma-rays about three times a second, beaming toward Earth. This pulsar is the first of its kind, one that may help scientists understand the nature of collapsed stars.

Antimatter is another mystery that has puzzled scientists for more than seven decades. Antimatter consists of antiparticles, elementary particles that have an opposite electrical charge as compared to the charge of ordinary particles. For example, while a proton has a positive charge, the antiproton has a negative charge. Moreover, when a particle of ordinary matter and a particle of antimatter come into contact with each other, they are annihilated, releasing huge amounts of energy.

According to the quantum theories developed to describe anti-particles, matter must be accompanied by an equal quantity of antimatter. The two are created in pairs out of pure energy. Thus, there should be an equal amount of both. Yet, all we see is ordinary matter.

So where is the antimatter? Well, one could imagine an entire planet, star, solar system, or even a galaxy made totally of antimatter. Theoretically, a planet made of antimatter would "appear" no different than some other planet. For this reason, astrophysicists have searched for antimatter by looking for evidence of violent activity that would occur at the boundaries of galaxies, where annihilations with ordinary matter would create abundant amounts of gamma-rays. But no such structures have been identified thus far. It appears we live in a world completely dominated by the ordinary matter we are most familiar. Yet, scientists believe that equal quantities of matter and antimatter were created during the Big Bang.

The search for antimatter began in 1930. It occurred right after Paul Dirac developed a mathematical equation that predicted the existence of an anti-world identical to ours but made out of antimatter. Paul Dirac (1902-1984) was a shy, quiet British theoretical physicist who derived the equation that describes the behavior of special particles and led to the prediction of the existence of antimatter. Dirac arrived at his astonishing result in 1928, using some of the concepts derived by the German physicist Max Planck (1858-1947) who founded quantum theory.

Planck proposed in 1900 that each light wave must come in a little packet, which he called a "quantum"—meaning that light has a dual nature; it behaves both as a particle and as a wave. By the 1920s, physicists began to apply the same concept to the atom and its constituents, resulting in the discovery of the quantum theory. Scientists soon found that quantum theory is not relativistic—the quantum description works only for particles moving slowly, and not for those moving at high velocity, close to the speed of light, and thus are called "relativistic."

Paul Dirac first derived an equation that combined quantum theory and special relativity to describe the behaviour of particles. Dirac's equation could have two solutions, in essence describing a particle with positive energy and a particle with negative energy. Dirac interpreted his unexpected result to mean that for every particle there is a corresponding antiparticle, exactly matching the particle but with opposite charge. For the electron, for instance, there should be an "anti-electron" identical in every way but with a positive electric charge. Dirac was right. A few years later the first antiparticles were discovered by American physicists Carl Anderson (1905-1991).

Anderson discovered antimatter in 1932, while working in a laboratory at the CalTech in California. He detected the track of a "positive electron," a particle as light as an electron but of opposite charge. Anderson's momentous discovery confirmed the antimatter theory and gave impetus to the search for the other antiparticles. Later,

Anderson named the anti-electron "the positron." The anti-protons were discovered in 1955. Since then, the search for antiparticles has driven major scientific breakthroughs in physics.

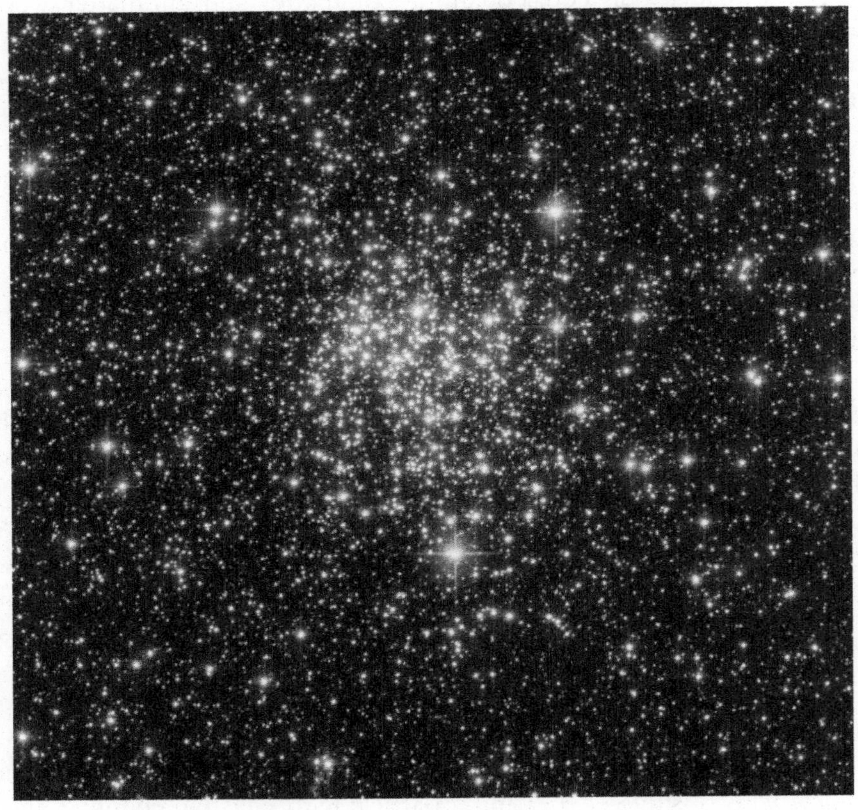

Star Cluster NGC 290, is an open cluster in the Small Magellanic Cloud, located in the constellation Tucana, about 200,000 light-years away and is roughly 65 light-years across. This cluster holds secrets of stellar evolution and more. Credit: European Space Agency & NASA. Acknowledgment: E. Olszewski (University of Arizona).

We still don't know why *our* Universe today contains matter but no antimatter. What happened to the antimatter? Of course, if matter and antimatter were created in equal amounts at the beginning, then we would not be here now. All matter would have annihilated with all antimatter.

In 1998, scientists at the CERN laboratory found a small asymmetry in the way particles and their antimatter equivalents transform into each other. Most recently, in 2010, researchers from FermiLab in Chicago reported that they discovered a new clue "that could help unravel one of the biggest mysteries of cosmology: why the Universe is composed of matter and not its evil-twin opposite, antimatter." If confirmed, the finding foretells fundamental discoveries at the new Large Hadron Collider, as well as a possible explanation for our own existence. These thrilling results may help solve the puzzle—why the world is mostly matter when the Big Bang should have created equal quantities of matter and antimatter. The problem of matter-antimatter asymmetry is not completely solved, though.

It is interesting to note that matter-antimatter annihilation—the complete conversion of matter into energy—could release the most energy per unit mass of any known reaction in physics. That's why engineers have considered antimatter as the propellant for a rocket that could power interplanetary and interstellar spaceships in the future. Propellant is any material in any state used to move or propel a rocket. In the antimatter rocket, the propellant consists of the antiparticles.

At its most basic level, an antimatter rocket can be viewed as a conventional rocket moving through action and reaction forces. For example, an anti-proton coming in contact with a proton yields energy. It also produces gamma-rays, leaving a spray of secondary particles that eventually decay into neutrinos and low-energy gamma-rays. In the antimatter rocket the gamma-rays would escape immediately. But the charged particles from the proton/anti-proton annihilation would produce the thrust force to propel the space vehicle to very high velocities.

The idea of an antimatter rocket is very attractive: a mixture of equal amounts of matter and antimatter would provide the highest energy density of any known propellant. Whereas the most efficient chemical reactions produce about 1×10^7 joules per kilogram (J/kg), the complete annihilation of matter and antimatter would produce 9×10^{16} J/kg, according to Einstein's mass-energy relation ($E = mc^2$). Theoretically, matter-antimatter annihilation would release almost ten billion times more energy than the hydrogen/oxygen chemical reaction that powers the Space Shuttle Main Engines, and 300 times more than the fusion reactions at the Sun's core.

Antimatter rockets, however, are not feasible at this time. Starting with the fact that antimatter does not exist in significant amounts in nature—at least not anywhere near the Solar System. Thus, it has to be produced in the laboratory. But it takes much more energy to make a tiny amount of antimatter than it would be needed for propulsion applications. Antimatter is created by energetic collisions in giant particle accelerators, such as those at FermiLab, in the USA, and at CERN, in Switzerland. The process typically involves accelerating protons to almost the speed of light and then smashing them with other particles. This process is complex and extremely expensive.

In addition, there are a number of design issues that must be worked out, some of which require technology not yet developed. The production and storing of antimatter, the containment and effective control of the antimatter-matter annihilation, the effective channeling of the charged particles to produce thrust, and other engineering design considerations are some of the most fundamental challenges for engineers to resolve. Antimatter rockets are very attractive as a concept, and many respectable scientists and engineers continue refining the idea and studying ways to overcome the design challenges. Yet, we have to wait a while before the technology is available to build an antimatter rocket.

But why would we need an antimatter-powered spaceship? Well, now that we know there are other worlds orbiting many stars in faraway places of the Universe, would not be interesting to entertain the notion of an interstellar voyage? Let's then talk about those extraterrestrial worlds and consider whether life may be possible there, too.

6
Are We Alone?

Life on Earth began over 2.7 billion years ago, right after the atmosphere stabilized. It is likely, thus, that life exists in other parts of the Galaxy once conditions are optimum there, too.

Messier 101, also known as the Pinwheel Galaxy, is a spectacular and nearby spiral galaxy located in Ursa Major. This beautiful galaxy is nearly twice the diameter of our own Milky Way Galaxy, and has a less prominent central bulge. Credit: NASA and ESA. Acknowledgment: K.D. Kuntz (GSFC), F. Bresolin (University of Hawaii), J. Trauger (JPL), J. Mould (NOAO), and Y.-H. Chu (University of Illinois, Urbana). Credit for CFHT Image: Canada-France-Hawaii Telescope/J.-C. Cuillandre/Coelum. Credit for NOAO Image: G. Jacoby, B. Bohannan, M. Hanna/ NOAO/AURA/NSF.

The possibility that there is life in other worlds outside the Solar System has transformed our view of the Universe and invigorated our impression that we are not alone. In just 15 years, astronomers have discovered many planets gravitating around other suns. It also makes us wonder whether some of those planets might support life. And by life I mean not just plants or simple organisms, but intelligent life like ours or even more advanced. Simply based on the enormous size of the Universe and the laws of probability, the odds are high that our planet is not the only inhabitable world.

Until the discovery of the first extrasolar planet in 1995, we knew almost nothing about the existence of other solar systems. Now the hunt for habitable, Earth-like planets that can support life has become one of the most thrilling scientific activities of this decade. Finding any form of life in another region of the Galaxy will be one of defining accomplishments of the 21st century.

Scientists estimate that at least 10 percent of stars similar to our own Sun have planets, which are known as *exoplanets* or *extrasolar planets* because they exist in planetary systems beyond our Solar System. The first exoplanet was discovered in 1995 using the radial velocity method at the Observatoire de Haute-Provence (a ground observatory in France). It was named 51 Pegasi b because it orbits 51 Pegasi a visible star in the constellation Pegaus, 50.9 light-years from Earth. Three years later, dozens more had been detected, and by 2008, the Hubble Space Telescope was able to take the first snapshot of the light reflected by another planet located 25 light-years away. The discovery—in such short time—of many types of planets circling other suns has opened another window to the heavens. As of July 2010, the count of known extrasolar planets stood at 473.

I am certain that the number of discoveries will continue multiplying in the next few decades, as the detection techniques are refined. Astronomers have already identified at least 350 stars in many parts of the Universe that are believed to possess planets.

Exoplanets are identified with the name of its host star followed by a lowercase letter, starting with "b" for the first planet found in a given star system (for example, 51 Pegasi b); "a" is skipped to avoid any confusion with the primary star. And so, for instance, exoplanet 55 Cancri f represents the fifth planet orbiting star 55 Cancri.

We are curious to know how similar the exoplanets are to the planets in our own Solar System. The majority of discovered exoplanets are giantic, resembling Jupiter in size but they are extremely hot because they gravitate very close to their host star. The exoplanets discovered to date are classified into several types including Terrestrial exoplanets, Hot Jupiters, Super Earths, and Hot Neptunes.

Terrestrial Exoplanets are similar to Earth or to any of the inner planets in our Solar System (closest to the Sun). A terrestrial planet in general is a rocky planet that is primarily composed of silicate rocks. Corot-7b is the first extrasolar terrestrial planet, discovered in 2009 by the satellite COROT. Planet Corot-7b orbits Corot-7, a G-type main sequence star, slightly smaller, and cooler than the Sun. The star, located in the Monoceros constellation, is even fainter than Proxima Centauri, and it has at least two planets. Corot-7b is first in many respects: It is the smallest known exoplanet; it is the closest exoplanet yet to its host star, which also makes it the fastest; it orbits its star at a speed of more than 750,000 kilometers per hour. Plus, data reveal the presence of a twin sister planet, another so-called super-Earth called CoRot-7c in this alien solar system. Corot-7b is the smallest exoplanet to have its diameter measured, at 1.7 times that of the Earth (which would give it a volume 4.9 times Earth's).

COROT (COnvection ROtation and planetary Transits) is the first robotic spacecraft dedicated to extrasolar planet detection. Launched in 2006, atop a Soyuz carrier rocket, COROT is a space mission led by the French Space Agency (CNES) in conjunction with the European Space Agency (ESA) and other international partners. Detecting the weak darkening or attenuation of the light emitted by stars during a transit event, the new planet hunter COROT detected the more exoplanetas, each one with his own peculiar characteristics, and also a brown dwarf.

Hot Jupiters are a class of giant extrasolar planet whose mass is close to or exceeds that of Jupiter (1.9×10^{27} kg). And because these huge gas exoplanets are so close to their stars their temperature is scorching-hot. That is the reason astronomers call them "hot Jupiters." While Jupiter orbits its parent star (the Sun) at 5.2 AU, hot Jupiters orbit between approximately 0.05 and 0.5 AU of their parent stars. Mercury, by comparison, orbits the Sun at about 0.37 AU. An example of a hot Jupiter is 51 Pegasi b, nicknamed Bellerophon, the first exoplanet discovered.

Super-Earths are extrasolar planets with a mass between that of Earth and the gas giants in our Solar System. The term super-Earth refers only to the mass of the planet and does not imply anything about the surface conditions or its habitability.

Hot Neptunes are extrasolar planets in an orbit close to its star (normally less than one astronomical unit away). The mass of a hot Neptune resembles the core and envelope mass of Uranus and Neptune. Recent observations have revealed a larger potential population of hot Neptunes than previously thought. The first hot Neptune planet was detected around Mu Arae (HD 160691). This is a main sequence G-type star approximately 50 light-years away from Earth in the constellation of Ara. The star has a planetary system with four known planets, three of them with masses comparable to that of Jupiter.

Most of the discovered extrasolar planets are gigantic planets, because they are easier to detect with available instruments and techniques. Predicting that hot Jupiters contain water vapor in their atmospheres, for example, astronomers searched for evidence and recently found it in an exoplanet named HD 189733b. The water it contains is too hot to condense into clouds. Nevertheless, previous observations of the planet from NASA's Spitzer Space Telescope and other ground and space-based observatories suggest that this exoplanet might have dry clouds, along with high winds and a hot, sun-facing side that is warmer than its dark side. Unlike Earth, planet HD 189733b is a gas giant, about 15 percent bigger than Jupiter and 30 times closer to its mother star than Earth is to the Sun. HD 189733b is located 63 light-years away in the constellation Vulpecula (the Fox).

A very interesting finding by Spitzer is the formation of an Earth-like planet around a binary star system called HD 113766. The system is composed of HD 113766 A and HD 113766 B, located 424 light-years from Earth in the direction of the constellation Centaurus. What makes HD 113766 A more interesting is that this star is very young, approximately 10 million years old (our Sun is abut 4.5 billion years old), is slightly more massive than the Sun, and there is a large belt of warm (~440 K) dust surrounding it. The dense dust belt, more than 100 times more massive than our own asteroid belt, is thought to be collapsing to form a rocky planet, which will lie within the star's terrestrial habitable zone where liquid water can exist on its surface. Studying HD 113766 A will provide more clues to help astronomers learn how rocky Earth-like planets form.

Exoplanets are so far from the Solar System and are so dim that they are difficult to detect; it's no wonder most planets found are giant, about the size of Jupiter or even bigger. Each planet discovered represents a valuable scientific accomplishment, but it has become much more exciting to search for terrestrial planets, those so named because they resemble Earth in size, and especially those in the habitable zone of their stars where liquid water and possibly life might exist.

The habitable zone (HZ) is defined as the region in a stellar-centered orbit where an Earth-like planet can maintain liquid water on its surface. The concept of the habitable zone arises from the assumption that water is essential for life. Water is needed as a solvent, and this process only occurs when water is in its liquid phase. And thus, water must exist in a planet for it to develop and sustain life like ours. As we know, of the nine planets in our Solar System, only Earth has liquid water in abundance. It is estimated that our planet holds a whopping 1,234,044,240,000,000,000,000 liters (3.26×10^{20} gallons) of water. Finding water or another liquid would confirm that some exoplanets may resemble Earth or other planets in our Solar System. In addition, for a planet to be habitable, it should also possess oxygen and organic compounds.

For our Solar System, the habitable zone extends from a point 5 percent closer to the Sun than Earth's orbit and up to a point 37 percent farther away from the Sun than Earth's orbit. If our planet were closer to the Sun, water would be boiling. If it were farther away from the Sun, water would be frozen.

Obviously, if the star is massive and luminous, the habitable zone is farther away compared to that for a low-mass, cool star. Astronomers believe that using the transit method to detect planets is advantageous, since the transit method allows them to calculate the size of the planet's orbit, and they can immediately tell if the newly discovered world lies in the habitable zone. They look for planets having a temperature of about 300 K, which would indicate a higher probability of having liquid water. They also examine the atmospheres of exoplanets for signs of life, particularly for the presence of certain life-related chemicals such as oxygen and carbon dioxide that would show in the color of the atmosphere.

Let's assume someone detects a green light around a distant exoplanet. The green color could be interpreted to indicate life of some form might be found on its surface. How so? Well, oxygen is created by plants as a by-product of photosynthesis, and oxygen produces a green atmosphere when stimulated by stellar subatomic particles. Even if a planet is not visible from Earth, astronomers are equipped with instruments to read the signature of the gases surrounding the planet. However, distinguishing the light from an extrasolar planet is a major technical challenge for astronomers, as the radiation emitted by its parent star nearby is overwhelming, making detection more difficult.

Star-Birth Clouds in M16: Stellar "Eggs" Emerge from Molecular Cloud.
Credit: NASA, ESA, STScI, J. Hester and P. Scowen (Arizona State Univ.)

The lightest exoplanet discovered so far is Gliese 581 e, which is just twice as massive as the Earth and orbits within the habitable zone of its parent star Gliese 581. This is a red dwarf star (small, with less than 40 percent the mass of the Sun and relatively cool star) located only 20.5 light-years away in the constellation Libra. It is the 87th closest known star system to the Sun. This star has at least four planets: Gliese 581 b, c, d, e. Exoplanet Gliese 581 c is also classified as terrestrial.

The star with the most exoplanets detected to date is 55 Cancri (55 Cnc). This is a binary star approximately 41 light-years away from Earth in the constellation of Cancer. The system consists of a yellow dwarf star and a smaller red dwarf star, separated by over 1,000 AUs (one thousand times the distance from the Earth to the Sun).

In 2009, five planets had been confirmed to be orbiting the primary star, 55 Cancri A (the yellow dwarf). The innermost planet is thought to be a terrestrial "super-Earth" planet, with a mass similar to Neptune, while the outermost planets are thought to be like the Jovian planets with masses similar to Jupiter. Star 55 Cancri A is ranked 63rd in the list of top 100 target stars for the NASA Terrestrial Planet Finder mission.

How do astronomers know that the distant cosmic object they perceive is a planet and not a star? Exoplanets are smaller and dimmer than their parent stars. And even with the best telescopes, those distant planets are difficult to detect. And so astronomers infer their existence using various scientific clues such as an excess of dust—indicative of planet formation—around the star, or use observatory instruments such as Doppler spectrographs and a number of other methods.

With the radial velocity technique, for example, the orbit size and mass of a planet are determined based on the perturbations it induces in its parent's star's orbit via gravity. Clearly, massive planets can be detected with this method, as their high mass perturb more its star, but to distinguish smaller, Earth-size planets scientists are now perfecting other techniques such as the transit method. The Hubble telescope has a special camera, called the Advanced Camera for Surveys coronograph, which also provides astronomers the ability to image some massive extrasolar planets.

In 2006, Alice Quillen of the University of Rochester predicted the existence of an exoplanet by looking at the debris disk of the star named Fomalhaut. She hypothesized that "the features...implied that there ought to be a planet, whose mass lay between that of Neptune and Saturn, orbiting nearby, some 119 astronomical units (AU) from Fomalhaut." She was right. Near the end of 2008, the Hubble Telescope detected the light reflected from that extrasolar planet and its coronograph camera took the first photo. Fomalhaut b, which has about three times the mass of Jupiter, orbits the bright southern star Fomalhaut, in the constellation Piscis Australis, or "Southern Fish."

The snapshot taken by the Hubble shows just a point of light that only a trained astronomer can identify as a planet, but it caused quite a stir.

Most methods of exoplanet detection use the effect of a planet's gravity on its star, which causes the star to "wobble" as the planet swings around it. The wobble may be detected by measuring the star's position relative to others, or (more commonly) by using the Doppler effect to measure the small back-and-forth change in the star's velocity. Both of these methods are effective for finding planets that have a very high mass, but are not effective at detecting smaller, Earth-mass planets, which have only a tiny effect on the motion of their parent star.

Scientists looking for small, Earth-like planets, use the transit method. With this technique, they look for a drop in the brightness of a star when a planet passes in front of it, an event which astronomers call a transit. This method will only detect those planets that cross the observer's line of sight from Earth to the star. But with enough sensitivity, the transit method is the best way to detect small, Earth-size planets, and has the advantage of giving both the planet's size (from the fraction of starlight blocked), as well as its orbit (from the period between transits). Since a transit only lasts a few hours, continuous monitoring is required.

In 2010, Hubble scientists reported that the hottest known planet in the Milky Way galaxy is being eating by its parent star. The giant exoplanet, called WASP-12b, is so close to its sun-like star (WASP-12) that it became superheated to nearly 2,800 degrees Fahrenheit and stretched into a football shape by enormous tidal forces. Its atmosphere has expanded to nearly three times Jupiter's radius and is spilling material onto the star. This happened because the planet got so hot that its atmosphere swelled so much that the star's gravity drew it in. The exoplanet, which is 40 percent more massive than Jupiter, may only have another 10 million years left before it is completely devoured. This effect of mass exchange between two stellar objects is commonly seen in close binary star systems, but this is the first time it has been seen so clearly for a planet-star system. WASP-12 is a yellow dwarf star located approximately 600 light-years away in the winter constellation Auriga.

Another peculiar small exoplanet—with 5 times the mass of the Earth—was found by Spanish astronomers in 2008. Identified as GJ 436c, this exoplanet orbits a dim red star (GJ 436) that lies at a distance 30 light-years away in the constellation Leo. The research team inferred the existence of GJ 436c by the way it disturbs the orbit of another planet in the system. Simulations showed that GJ 436c orbits its host star in only 5.2 Earth days, and it appears to complete a revolution in 4.2 Earth days, compared to the Earth's revolution of 24 hours and full orbit of 365 days.

In September 2007, astronomers identified the oldest exoplanet yet discovered, a planet orbiting a dying star located about 4,500 light-years away from the Solar System. The planet, named V391 Pegasi b, has a surface temperature of 200 degrees Celsius. This is a rather interesting exoplanet to study because it will help astronomers understand better what could happen to our own planet Earth in the distant future.

Exoplanet V 391 Pegasi b has survived many cataclysmic events as its parent sun ends its life. Observing the changes of this exoplanet while its mother star is dying will give researchers a preliminary picture of what could be Earth's destiny in four to five billion years. That's when our own Sun will exhaust its hydrogen fuel, expand enormously as a red giant and will expel its outer layers in an explosive helium flash. Under similar cosmic calamity, the fate of our planet Earth may be the same to that of V391 Pegasi b.

Recently, astronomers discovered a bizarre planetary system where not all planets orbit their host star in the same plane. We are accustomed to our elliptically-shaped Solar System, with its eight relatively solitary planets whose orbits are almost circular and lie within a nearly-flat disc called the ecliptic plane. But this neat arrangement is not the same in other planetary systems. Astronomers first detected three Jupiter-type planets orbiting the yellow-white star Upsilon Andromedae. Recently they discovered that the orbits of two of the planets are inclined by 30 degrees with respect to each other. Such a strange orientation has never seen before in any other planetary system.

This finding surprised the team of astronomers, but, why should we expect that everything in the Universe should be just as we observe in our cosmic neighborhood? If not all stars are the same, why should their planets? Theories of how planetary systems form and evolve must accommodate many types of events that could disrupt planets' orbits after a planetary system forms. I am convinced that many more unexpected discoveries in the next years will change our perception of the Universe.

The search for habitable exoplanets is probably the most exciting activity in astrophysics today. NASA's Kepler spacecraft was launched in February 2009 to search for signs of Earth-sized and smaller planets, concentrating its synthetic eyes on 100,000 solar-like stars found 3,000 light-years away from us. Kepler, known as a planet-hunter spacecraft, is scheduled for a 4-year mission to help scientists select the places in the sky for future searches.

Early in 2010 members of the Kepler science team announced the discovery of the first five new exoplanets. The discoveries show that the instrument is working well and, although the planets are not Earth like but rather they are huge hot Jupiters (2200 to 3000 degrees Fahrenheit, hotter than molten lava and much too hot for life as we know it), I am convinced that very soon extrasolar planet hunters may surprise us with the news that they have spotted one planet that looks just like our Earth. Imagine that!

Kepler is designed to look for the signatures of planets by measuring dips in the brightness of stars. As we said earlier, when planets cross in front of, or transit, their stars as seen from Earth, they periodically block the starlight. The size of the planet can be derived from the size of the dip. The temperature can be estimated from the characteristics of the star it orbits and the planet's orbital period.

Kepler could also detect exo-moons! Yes, Kepler's high definition data may produce sufficient detail to detect moons orbiting an exoplanet. In fact, the exoplanet TrES-2b (the most closely studied planet for the possible presence of a natural satellite) is in the field of view of Kepler, and scientists eagerly analyze the data hoping to discover exo-moons smaller than Earth.

Planet TrES-2b, also known as Kepler-1b, orbits the star GSC 03549-02811 located 750 light-years away. It is a gas giant a bit bigger than Jupiter, but TrES-2b is located very close to its star, making it a hot Jupiter exoplanet. Being so large the likelihood of TrES-2b possessing at least one moon is very high.

On the other hand, having discovered Earth-size extrasolar planets gives researchers hope for finding alien life. Of course the size of the planet is not what would determine the possibility of life. There may be many other factors, aside from the habitability conditions—where an Earth-like planet can maintain liquid water on its surface, or be capable of bearing extraterrestrial life similar to ours. Can intelligent beings survive in other conditions?

This image zooms into a small portion of Kepler's full field of view—a 100-square-degree patch of sky in our Milky Way galaxy. At the center of the field is the GSC 03549-02811star with exoplanet TrES-2 zipping closely around it every 2.5 days. Image credit: NASA/Ames/JPL-Caltech.

We know that it took about 4600 million years for intelligent life to evolve on the Earth. Will it be a similar process elsewhere in the cosmos? Moreover, should we also consider the possibility that life in other parts of the Universe exists in forms different from those on Earth? Certainly! We must expand the search for extraterrestrial life to include what scientists call "weird life," that is, organisms that lack DNA or other life molecules found on Earth. Researchers use what we know about life on Earth to guide their search for life on other extraterrestrial worlds, and as such they have focused their research on detecting water. But there could be other kinds of chemistry that support life, in a form that would differ from life as we know it.

Instead of water, can life exist in other liquids such as ammonia or methane? Maybe. What about the hereditary material in the human race, does DNA have to be the same? DNA uses phosphorus in its backbone. Is it possible to build a backbone out of another element instead? At the same time we must consider that, if life exists in other planets, it may not have the same metabolism based on carbon like ours, and thus detecting its presence may require more technical sophistication. A group of Argentinean scientists announced recently that they have developed a method to detect any form of life. Their concept still needs further development, but their work represents another step in development of technology needed to detect life in alien worlds.

Many ideas as yet formulated should not be excluded, as they could be within the realm of possibility, since extraterrestrial life may be so different that we have not imagined it.

Before we leave this chapter, let me ask one more question: *Could the seeds of life on Earth have come from another planet?* Assuming there is life somewhere else in the Galaxy, could microbes have survived sealed deep inside the millions of rocks hurled across the Solar System? The Transpermia Theory states that life on Earth originated in other parts of the Universe.

Transpermia studies a possible mechanism for transfer of life between planets via rocks ejected by major asteroid or comet impacts. The term "transpermia" was coined by Oliver Morton (a British science writer) to describe the transfer of lifeforms by this method, and to distinguish it from the more general concept of panspermia. Possibilities for transpermia include hypothetical Mars-life reaching Earth; Earth-life reaching Mars, the Earth's Moon and other moons of the outer Solar System; and interstellar transfers via meteoroids. Transpermia is a hotly debated concept in the space exploration community. That is why we need to test it and probe one way or another if this idea has any merit.

To test the Transpermia theory scientists have carefully planned a sample return mission to Phobos, the largest moon of the planet Mars. The mission known as Phobos LIFE (Living Interplanetary Flight Experiment) will use a Russian rocket to lift a spacecraft and send it off toward the Martian moon. It will carry onboard a small canister with 30 samples of micro-organisms to test the Transpermia theory by determining whether microbial life from Mars has made it to Earth. The Phobos sample return mission will travel to Phobos and back over a three-year period. If any microbes survive transit through deep space, it could change what we think about the origins of life on our beloved planet.

Microbial life could also exist in Europa, the icy moon of Jupiter. A new theory states that an ocean on that moon is being heated by gravitational forces stretching its core as it moves around the giant Jupiter. Some scientists are convinced that the shell of ice on Europa is thinner than previously predicted. It is then possible for some form of life to exist beneath its thin crust. Other planets and moons could also harbor life. And if results from the Phobos experiment are positive, we could believe that the seeds of life on Earth indeed came from other parts of the Solar System or from other regions of the Galaxy. The crucial question to answer is whether the tiny life forms could make it to our planet.

Now let us turn our attention to the scary beasts that roam the Universe, the cosmic dragons that threaten and devour the pretty stars that live near them.

7
Dragons in the Sky

Black holes, cosmic monsters that devour stars around them and spit out deadly energy

Circinus Galaxy Spews Hot Gas into Space. This is a Type II Seyfert galaxy and closest known active galaxy to the Milky Way, only 15 million light-years distant. At the center is an active galactic nucleus, where matter glows brightly before likely spiraling into a massive black hole. Credits: NASA, Andrew S. Wilson (University of Maryland); Patrick L. Shopbell (Caltech); Chris Simpson (Subaru Telescope); Thaisa Storchi-Bergmann and F. K. B. Barbosa (UFRGS, Brazil); and Martin J. Ward (University of Leicester, U.K.).

Imagine you are sailing in deep space, far, far from the Earth and the Sun, awestruck by the splendor of the stars shining against the blackness of the cosmic sea. Suddenly, your spaceship is suctioned by a gigantic force much greater than the power of the rockets and is impossible to overcome. Could this be the colossal gravitational pull of a black hole? If so, then you are doomed, as it would be utterly impossible to resist it.

A black hole is a cosmic object so massive that its gravitational field is powerfully intense to cut off a region of space from the rest of the Universe—no matter or radiation (including visible light) that enters the region of influence of the black hole can ever escape. The massive object of extraordinary density is surrounded by a spherical border, called the event horizon, through which matter can enter but it cannot exit, and for that the object is called black.

The size of the event horizon is determined with the Schwarzschild radius, which is defined mathematically with the relation $r_S = \frac{2G \cdot M}{c^2}$, where G is the universal gravitational constant, M is the mass of the object, and c is the speed of light. The universal constant G is equal to 6.67×10^{-11} N m^2 kg^{-2}, and is the same in Newton's law of universal gravitation that describes the gravitational attraction between bodies with mass. The definition of the gravitational law is found in the Glossary.

Since G and c are constants, we solve $r_S = 1.48 \times 10^{-27} M$ (m/kg). This simple formula states that the gravitational field of a Schwarzschild black hole depends on one parameter: a mass M. Thus, for a sphere with a solar mass ($M = 1.98892 \times 10^{30}$ kg), the event horizon has a radius of only 2.94 km. The smallest stellar black hole discovered so far has a radius of about 15 km (9.3 miles). An average stellar black hole of about 10 solar masses has a radius of about 30 km (18.6 miles), while a big stellar black hole may have a radius of up to 300 km (186 miles).

The radius of the event horizon is named after the German astrophysicist Karl Schwarzschild (1873-1916) who derived the relation. A child prodigy who published his first scientific paper on celestial orbits at age 16, Schwarzschild was a gifted mathematician. He is better known for finding the first exact solution to Einstein's general equations of gravitation just months after Einstein published them in 1915. Schwarzschild died shortly after. He did not know it then, but his work led to a description of how a mass curves space and provided the fundamental insight leading to the study and further discovery of black holes.

There are two main types of black holes: supermassive and stellar. A *stellar black hole* results from the collapsed remains of an exploded supergiant star. The core of the star that remains cools down, as nuclear reactions are no longer taking place to produce heat. This reduces the pressure pushing outward from within the star. Gravity pulls the material of the remnant star toward its center. If the force of gravity is strong enough, the core continues to shrink until is no longer visible. The core becomes so small and super dense that its resulting gravitational pull would be enormous. The supermassive but tiny stellar corpse then becomes a black hole.

Stellar black holes may be common in parts of the Universe that contain young cosmic structures. I assume this because, due to their extreme masses, supergiant stars have relatively short lives, only 10 to 50 million years, and are mainly found in open clusters, the arms of spiral galaxies, and in irregular galaxies.

On the other hand, a *supermassive black hole* is one with mass equivalent to the mass of billions suns. This type of black hole exists in the centers of galaxies, including our own Milky Way. The exact origin of supermassive black holes is not yet clear, but they are believed to be the byproduct of the process of galaxy formation. It is also possible that two—not just one—supermassive black holes can co-exist in the same galaxy, as it was discovered in 2002. Astronomers using the Chandra X-ray telescope found the first evidence that, in a galaxy 400 million light-years away, two immense black holes are drifting toward each other. They estimated that, in a million years, the black holes will merge violently with an eruption of energy and a burst of gravitational waves that could warp the fabric of space.

And there is a third type of black hole known as a microquasar. Microquasars are stellar black holes in our Galaxy that mimic, on a smaller scale, many of the phenomena observed in quasars. A quasar is a star-like extragalactic object, a source that emits huge amounts of electromagnetic energy in the form of light and radio waves, making it among the most luminous and thought to be the most distant objects in the Universe. Discovered less than 20 years ago, microquasars are believed to be powered by spinning black holes and with masses of up to a few tens times that of the Sun.

The idea of a star becoming a "black hole" can be traced to the eighteenth century. Combining Newton's theory of gravitation with his conclusion that light is made of particles (corpuscular light theory), scholars began to study the effect of gravity on light. French mathematician Pierre-Simon Laplace (1749-1827) introduced the idea of the "black star," considered the predecessor of the black hole concept.

In a 1798 treatise, Laplace reasoned that if enough mass were added to a star like the Sun, the gravitational force of the star eventually would become so great that the light particles would not be able to leave the star. Laplace concluded: "it is therefore possible that the greatest luminous bodies in the universe are on this account invisible." Laplace believed that because light could not escape from the gravitational pull, the star would not be visible to an observer against the dark night sky. The invisible star would be in fact a *black star.*

But this idea was not widely accepted, especially since another respected scientist determined that light was a wave, not made of particles as Newton had postulated. Thus, it was not until Einstein stated the theory of general relativity that the modern concept of black hole was conceived. By then, the duality of light was an accepted fact. Einstein's analysis of the photoelectric effect in 1905 demonstrated that light also possesses particle-like properties, and this was further confirmed with the discovery of the Compton scattering in 1923.

Einstein's theory predicts that when a large enough amount of mass is present within a sufficiently small region of space, all paths through space are warped inwards towards the center of the volume. When an object is compressed enough for this to occur, collapse is unavoidable and a black hole is formed. In this condition, it would take infinite strength to resist collapsing the mass into a black hole.

When any object crosses the event horizon and falls into the black hole, it will vanish without a trace. To break away from a black hole would require a speed greater than that of light, c, which is not allowed. Thus, the object reduces to the singularity of the black hole. Mathematically, a singularity implies the solutions to the equations are undefined. Physically, a gravitational or spacetime singularity is, approximately, a place where quantities used to measure the gravitational field become infinite.

Considering all that, what would happen to a spaceship crossing the event horizon and falling into a black hole? Well, I will not attempt to describe such terrifying scenario. However, be warned. Once a body crosses the event horizon, there is no way to turn back; it will be lost forever within the black hole. It is said that black hole is a tiny region of space that contains the end of time. If it were to fall into such hole, the spaceship would find the end of the Universe, in fact the end of everything!

What is even scarier is the fact that black holes eat stars. It's true. If material from a nearby star crosses the event horizon, the black hole will gobble it up. During the process of ingestion, gas and matter spiral inward, heating up to very high temperatures and emitting large amounts of light, X-rays and gamma-rays. The 2006 NASA/ESA image (below) shows a super-massive black hole ripping apart a star and devouring a huge portion of it. This astronomical event was confirmed several years ago by NASA's Chandra and the European Space Agency's XMM-Newton X-ray Observatories.

Astronomers believe a doomed star came too close to a giant black hole after being thrown off course by a close encounter with another star. As it neared the enormous gravity of the black hole, the star was stretched by tidal forces until it was torn apart. Credit: NASA/CXC/M.Weiss (2006).

The term "black hole" was first uttered during a lecture in 1967. Thirty years earlier, the famous American theoretical physicist J. Robert Oppenheimer (best known as "the Father of the Atomic Bomb") and his student Hartland Synder published in an article the first description of a black hole but without naming it so. The abstract of their article started with "When all thermonuclear sources of energy are exhausted a sufficiently heavy star will collapse." Oppenheimer and Snyder were referring to a massive object that collapsed and they defined it as a mathematical singularity because of its infinite density. The article did not attract much attention, perhaps because it was published on the same day the world war started. Or perhaps because at that time researchers were not excited about black holes since a real cosmic black hole had not been discovered.

Then, during a conference in 1967, the eminent American physicist John Archibald Wheeler referred to Oppenheimer-Synder's results. During his presentation he used the expression an "object that completely collapses gravitationally," saying so often that a member of the audience suggested to just call it "black hole." It seems that Wheeler agreed because at the end of the same year the term "astrophysical black hole" appeared for the first time in print in an article he published. Thus, even though he is credited with coining the word, Wheeler himself insisted that it was suggested to him by somebody else. John Wheeler, the "father of black holes," died on April 13, 2008. He was 96.

The theory of general relativity describes a black hole as a region of empty space with a singularity at the center. This description changes when the effects of quantum mechanics are taken into account. In 1975, Stephen Hawking published a shocking idea: that if one takes quantum theory into account "black holes are not quite black." In other words, black holes can radiate energy, glowing slightly with radiation consisting of photons, neutrinos, and other particles. Eventually, black holes would radiate all their energy and vanish.

Hawking published these results after the Mexican-Israeli physicist Jacob Bekenstein put forward that black holes have thermal properties and thus should radiate energy. The thermal radiation thought to be emitted by black holes due to quantum effects is known by some as Bekenstein-Hawking radiation or most commonly just Hawking radiation. Despite its strong theoretical foundation, this radiation has never been observed.

Although invisible to the naked eye, astronomers can find black holes by detecting their gravitational influence on the neighbor stars. Some researchers believe that massive black holes may be roaming our Galaxy, hidden inside strange star clusters. They theorize that the Milky Way formed from smaller galaxies smashing together, some of which could have had super-massive black holes at their cores. If they are correct, the black holes should still be floating around somewhere.

Imagine that! The powerful gravity of all those black holes would influence hundreds or thousands of stars around them.

The mass of a black hole depends on the age of the galaxy where it is located. Older galaxies have black holes that are relatively calm. However, younger galaxies tend to be very active and super-massive black holes at their cores would consume more matter and produce abundant X-rays in the process. And since black holes grow by attracting and ingesting matter from stars around them, I wonder how long it would take for a stellar black hole to get fat and become a super-massive black hole.

In 2010, an astronomer spotted a super-massive black hole—heavier than one billion suns—headed out of its home galaxy at more than a million kilometers per hour (1.08×10^6 km/h or 670,000 mph). The titanic black hole, located thousands of light-years from the center of its galaxy, may have resulted from the collision of two holes. When astronomers observed (with the Hubble Telescope) a bright spot, they concluded that because of how strong the emissions were, the spot could not be a star. Their explanation is that more likely two spinning, super-sized black holes collided in the center of the galaxy. The spin of the black holes resulted into a kick able to propel the heavyweight merged hole from its perch at the center of a galaxy.

It is quite intriguing to me the idea that inside a black hole, space and time switch roles. Some researchers theorize that in a black hole there is a zone where space becomes time, and time becomes space. This would be due to the extreme warping of spacetime in the immediate vicinity of the singularity—the frozen, collapsed, ultradense point where the density of matter and the gravitational field are infinite. Other scientists argue that this idea makes no sense. Who knows? A lot more research is needed to confirm that.

Other groups of physicists believe that the final, correct description of black holes may require a theory of quantum gravity. Quantum gravity is the field of theoretical physics that attempts to unify quantum mechanics with general relativity. It is one of the most challenging problems in science right now that requires advances in mathematics and physics to resolve it. If you have a strong intellect, and like scientific dares, this may be an exciting study to pursue.

Interestingly, researchers from around the world are working at CERN to restart the supercollider with the experiments that will simulate the birth of the Universe. Some people believe that the attempt to re-create energies near those of the Big Bang would create miniature black holes that would swallow the Earth. I don't believe it. I am convinced that the world will not end when the LHC supercollider produces the expected proton collisions that will create a microscopic black hole. According to the Hawking radiation theory, a black hole could radiate energy and the smaller the black hole, the more intense the radiation would be.

I should think that the miniscule black hole in the CERN supercollider would evaporate very quickly and vanish almost instantaneously. And even if the black hole survives for a microsecond, it would have a tiny mass (mass of a hundred or so protons) and be less than a thousandth the size of a proton; thus, its gravitational pull would be exceedingly weak and would not be able to swallow any matter at all.

If you are reading this book it means I am correct, because I wrote these words much before the CERN supercollider was put to work. I am so confident the world will not end that now I close this chapter to write about another topic needed to continue exploring the possibilities of interstellar travel.

8
Wandering among the Stars

If I could exceed the speed of light, would I reach my future instantaneously?

The Cygnus Loop Supernova Remnant. 15,000 years ago a star in the constellation of Cygnus exploded. This image shows a small portion of the Cygnus Loop supernova remnant, taken with the Wide Field and Planetary Camera on NASA's Hubble Space Telescope on April 24, 1991. The Cygnus Loop marks the edge of the expanding blast wave from that colossal stellar explosion. Credit: J.J. Hester (Arizona State University), and NASA. Co-investigators: P.A. Scowen (Arizona State University), Ed Groth (Princeton University), Tod Lauer (NOAO), and the WFPC Instrument Definition Team.

I am fascinated by the idea of traveling faster than the speed of light. And I imagine that we can move across the Universe, leaping from galaxy to galaxy, from star to star, seeing up-close all there is between them. No wonder science fiction novels and films are so popular. In the movies, awesome large spaceships zip around space carrying onboard an entire community of people and everything they need to live for years and years, making interstellar travel seem so easy. In real life, however, this scenario is impossible—at least at this time. If we know and understand the laws of physics, we are faced by the realization that no spaceship can travel faster than the speed of light, the speed that would be required to cross a small fraction of the Universe.

Einstein's theory of relativity is the foundation of modern physics, and the assumption that the speed of light is constant and finite is a formidable cornerstone. The speed of light, thus, is *the* cosmic speed limit. Although it is very fast compared to the speed at which we move here on Earth or to the speed of our space vehicles, the speed of light is slow compared to the size of the Universe—there is a huge scale disparity, especially when we consider the span of human life. In the scale of cosmic time, we live for only a fraction of an instant. Eighty or even one hundred years is an insignificant time span compared with the time required to cross the Galaxy at the speed of light. As you may remember, the Milky Way is 100,000 light-years across.

Einstein's special relativity establishes two postulates that form the foundation for our understanding of the cosmos. One of these claims is that *nothing can travel faster than the speed of light.* And so, limited to a speed equal to that of light, a spaceship would take 4.3 light-years to reach the nearest star (beyond the Sun). But for an object to approach the speed of light, an infinite amount of energy is required. Therefore, at this time, the idea of interstellar travel is terribly difficult to conceptualize without violating the laws of physics.

What if the laws of physics are different when the scale of spacetime changes? Then we would modify our thinking and conceive new ideas for interstellar travel. We know that light doesn't travel far enough in one year. Not far enough with respect to the scale of the Universe. Thus, in order to make it across, we must distort regions of spacetime and shorten the path, right?

Faster than light travel (FTL) is an intriguing prospect. The concept of FTL implies that matter (and information) can be propagated at speeds faster than the speed of light. However, true FTL, in which matter exceeds the speed of light in its own local region, is impossible, at least for a spacecraft. Therefore, without a way to break or circumvent the speed of light, for humans to travel outside of our Solar System would be problematic if not impossible. It's true that special relativity does not forbid the existence of particles that move faster than light. Nevertheless, the existence of such hypothetical particles—called tachyons—has neither been proven nor disproved. But that's another story altogether. Here we are only interested in answering whether a spaceship can travel at the speed of light or faster.

Several good ideas for interstellar travel have been conceived that do not violate the laws of physics. Two ideas discussed in the scientific literature are space warp drives and wormholes. A space drive is a form of propulsion where the fundamental properties of mass and spacetime are used to create propulsive forces in space without carrying or throwing a reaction mass, like in conventional rocket engines. A wormhole is a hypothetical tunnel through spacetime.

These ideas arise from what physicists refer to as "apparent" or "effective" FTL. Apparent FTL assumes that unusually distorted regions of spacetime might permit matter to reach distant locations faster than light, taking the "normal" route (though not faster than light moving through the distorted region).

Miguel Alcubierre, a Mexican theoretical physicist, conceived a scientifically sound warp drive concept. In 1994, Alcubierre published a paper in which he described a way for an object to move at speeds greater than the speed of light. The basic idea behind the Alcubierre warp drive is to expand spacetime behind the spaceship, and contract it in front of it. Unfortunately, warping spacetime would require negative energy, something that we don't have.

Another theoretical idea of apparent FTL is the traversable wormhole. Wormholes are the tunnels or superhighways through spacetime that might connect two different regions of the Universe. If the tunnel is wide enough, a body could enter the wormhole at one end and emerge instantaneously at the other, at a different time, a different place, or even in another universe entirely! In any case, a spaceship could travel a colossal distance through the tunnel in just minutes.

We can imagine that spacetime is distorted in such a way that the path would be shrunken for a spaceship to cross it in less time. For example, if you were to enter one end of a wormhole, say at the edge of our Solar System, you could (in theory) emerge from the other side an arbitrary distance from where you entered, on the opposite side of our galaxy perhaps. It would appear as if you traveled faster than light—though you didn't.

Wormholes exist as a mathematical solution to Einstein's field equations. The idea originated in 1915 when Einstein predicted that gravity distorts space (according to his theory of general relativity). In the 1930s, Einstein and his colleague Nathan Rosen (1909-1995) concluded that a black hole might connect to another through a tunnel through spacetime. This idea is known as the Einstein-Rosen Bridge.

It has been suggested that a spaceship could fold a space-time bubble around itself to travel faster than the speed of light. This idea involves manipulating dark energy to propel the vehicle forward without breaking the laws of physics. In 2010, Stephen Hawking proclaimed that humans could travel into the future in spaceships flying at higher than light speed velocities. Of course, we cannot contradict him. But without the energy required to propel a rocketship to those hyper or superluminal speeds, we just cannot hop from galaxy to galaxy, not even from star to star.

Space drives and wormholes are theoretically possible, but both would require extraordinary advancements in technology not foreseeable in the near future. Although apparent FTL is not excluded by the theory of general relativity, the physical plausibility of these solutions is uncertain. It's true that ultra dense matter can bend space, but to bend space enough to get light-years away quickly takes infinite amounts of energy. I don't know what kind of technology would be required to warp space. Try bending just the small space around you, let alone stretches of space between stars, can you even imagine it? Perhaps someday someone will figure out how.

Propulsion is the key technology that will allow us to reach the stars. Reaching the next space frontier requires a broad expansion of advanced propulsion ideas and technologies. The ultimate goal of our space exploration dreams is not just to reach the planets, but to travel out of the Solar System. Although it is not impossible, it is so difficult that there is not one clearly feasible approach that I can identify now.

To date, viable or not, all ideas of FTL are outside our technology reach. Yet, I hope one day someone finds a way to bend space or can develop propulsion technology for people to travel at near light speed. For now I have no option but conclude that intergalactic travel remains for now a topic of research and scientific inquiry.

9
Space Robots

Exploring the heavens to unravel its mysteries

An artist's impression of the Voyager spacecraft. Image Credit: NASA

Since the historic launch of the first Sputnik satellite in 1957, many robotic spacecraft have left Earth behind and moved off into deep space to explore it on our behalf. The space probes, carrying advanced instruments but no people, have surveyed the heavens and given us the most splendorous views of the planets. Robots have journeyed across the Solar System, going to the planets and their moons, to comets and asteroids, and have probed our mother star. Space robots have gone where no human being has gone before and revealed many secrets of the Universe.

Depending on the type of operation they perform, space robots are of three types: fly-by, orbiter, and lander. A fly-by spacecarft operates in outer space, and as it passes a planet or moon, it images the landscape and makes observations of temperature, radiation, and other characteristics of the planet's environment. An orbiter is a probe that orbits around a planet, just like a satellite, and performs similar activities as a fly-by probe but its mission around a planet is much longer. NASA's Lunar Crater Observation and Sensing Satellite (LCROSS) is a typical example of an orbiter space probe that is taking data of the Moon and recently confirmed that there is water in a crater. And finally a lander is a robot that lands on a planet to perform experiments on its surrounding, releasing chemicals, or digging into surface material to examine and test its composition.

There are several types of landers: impactor, hard landers, soft landers, and penetrators. Impact spacecraft, as the name implies, crash on the surface of a planet or moon on a controlled manner, for example the Japanese Kaguya impact probe directed in a controlled impact against the Moon, and the Chandrayaan-1 lunar probe from India. Hard landers are spacecraft that have special systems to cushion the impact while landing. Soft landers, on the other hand, touch down gently, aided by special thrustable rockets and/or parachutes. Finally the penetrator landers are designed to penetrate the surface to extract samples, for example.

Finally, depending on its method of operation, probes are of two types. Sample-return probes are spacecraft that bring back material samples to Earth, such as Stardust and Hayabusa that collected samples from a comet and an asteroid, respectively. NASA's Stardust spacecraft passed by Comet Wild 2 and collected dust samples from the comet's coma while imaging the nucleus; it returned to Earth January 15, 2006. The Japanese Hayabusa probe returned to Earth in June 2010 after a rendezvous with asteroid 25143 Itokawa. And there probes that perform necessary operations on the planet or in space and does not return, such as the Pioneer and Voyager spacecraft the most intrepid robots, that now move quietly in the cold, dark interstellar space, navigating towards the distant stars.

The twin Pioneer and the twin Voyager are the first interstellar spaceships ever built by the natives of Earth. Launched in 1972, Pioneer 10 was the first fly-by space probe to study Jupiter and its environment. Jupiter is the fifth planet from the Sun (390,682,810 miles away at the closest point in our orbit), and it is the largest in the

Solar System. Flying at 82,000 mph, Pioneer 10 reached its destination on December 3, 1973 and obtained the first images of the planet.

In April 1973, NASA launched the twin probe Pioneer 11, the first vehicle to fly-by Saturn, its rings and moons. After acquiring additional gravitational energy from Jupiter, Pioneer 11 followed an escape trajectory from the Solar System. This maneuver, known as "gravity assist," gives a spacecraft a small fraction of the planet's orbital energy to help it change direction and speed without using its own propellant. When the last signal was received in 1995, the robot was at a distance of 44.7 AU from the Sun. Pioneer 11 is now headed toward the constellation Aquila (The Eagle), northwest of the Sagittarius constellation. If it would not disintegrate by cosmic dust or other interstellar effects, it would take about 4 million years for Pioneer 11 to reach the nearest stars in that group.

An artist's impression of the Pioneer spacecraft. Image Credit: NASA

The Pioneer spacecraft went on their long journey carrying onboard a gift for the extraterrestrial beings they may encounter—if they exist. Pioneer carried a gold plaque with a drawing of a man and a woman, showing Earth's location in the Solar System and our place in the Galaxy. The drawing is engraved into a gold-anodized aluminum plate, 152 by 229 millimeters (6 by 9 inches), attached to the spacecraft's antenna support struts in a position to help shield it from erosion by interstellar dust.

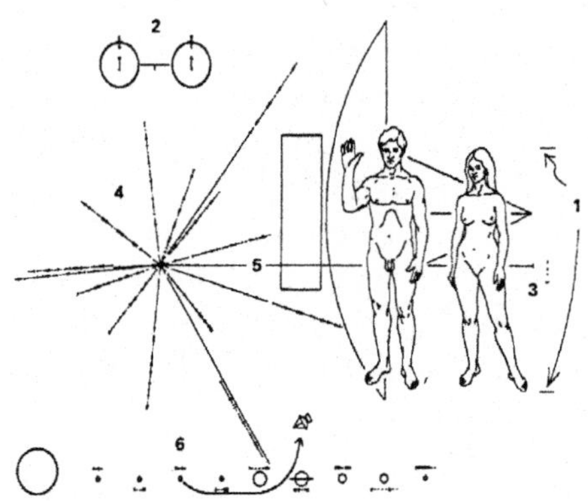

Pioneer Plaque Symbology. Image Credit: NASA

After launch, Pioneer 10 reached a speed of 52,140 km/h (32,400 mph) needed for the flight to Jupiter, making it the first fastest human-made object to leave the Earth; it was fast enough to cross Mars' orbit in just 12 weeks. On July 15, 1972, Pioneer 10 entered the asteroid belt. This is a wide region located between the orbits of Mars and Jupiter that contains millions of asteroids ranging widely in size from about 940 km in diameter (one-quarter the diameter of our Moon) to bodies that are less than 1 km across.

Accelerating to a speed of 131,963 km/h (82,000 mph), Pioneer 10 passed by Jupiter on December 3, 1973; its mission officially ended on March 31, 1997, after exploring the outer regions of the Solar System. Its sister ship, Pioneer 11, ended its mission September 30, 1995, when its last transmission was received.

Pioneer 10 sent its last signal to Earth, a final farewell before leaving us forever after a 30 year travel through the Solar System. Its last, very weak signal was received on Jan. 22, 2003. The robot was about 82 times the nominal distance between the Sun and the Earth.

At that distance, it takes approximately 11 hours and 20 minutes for a radio signal to reach the Earth. The spaceship now drifts away in interstellar space, heading in the direction of the red star Aldebaran in the constellation Taurus (the Bull). Aldebaran is about 68 light-years away. To reach that star, Pioneer 10 would have to maintain its current speed for more than two million years!

Two other twin spacecraft named Voyager 1 and 2 were launched in 1977 on a mission to explore the outer Solar System. The probes visited Jupiter, Saturn, Uranus, Neptune, and 48 of their moons, and studied the planets' unique system of rings and magnetic fields. The Voyager also made many intriguing discoveries, including a possible ocean of frozen water on one of Jupiter's moons. Voyager 2 is the only spacecraft ever to have visited the gas giants Uranus and Neptune. Voyager 1 overcame Pioneer 10 and became the most distant human-made object, traveling through the heliosphere and is now heading into interstellar space.

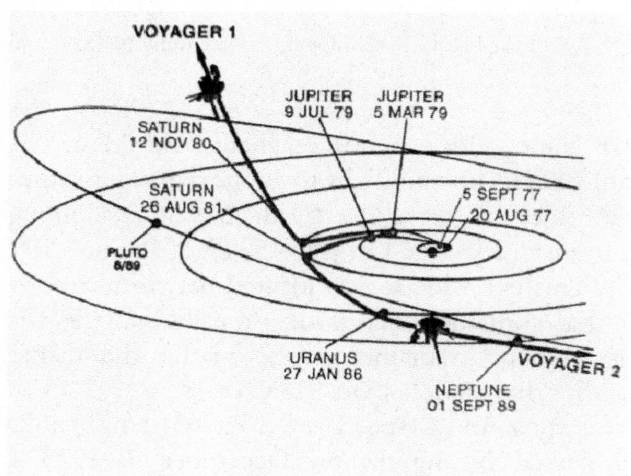

Trajectory of the Voyager Spacecraft. Credit: NASA

The heliosphere is a bubble of magnetism from the Sun, inflated to gigantic proportions by the solar wind. The bubble extends beyond the orbit of Pluto, to about 100 UA, making it the boundary that separates our Solar System from interstellar space. The heliosphere is very important for us because it deflects galactic cosmic rays, those high-energy particles from black holes and supernovae that otherwise would penetrate the Solar System and harm life on the Earth.

The Voyager also sailed away bearing gifts. The spacecraft carried along "*Murmurs from Earth*," a phonograph record—a 12-inch gold-plated copper disk—containing sounds and images representing the rich diversity of life and culture on Earth. NASA scientists recorded 115 images of our beautiful planet and a variety of its natural sounds, adding music from different cultures, plus spoken greetings in fifty-five languages. This record is a gift from the people of our beautiful planet to any extraterrestrial being who may encounter the robots along their cosmic travel.

Interestingly, none of the four interstellar spacecraft is headed towards Proxima Centauri, our closest neighbor star beyond the Sun. If the Voyager were moving in that direction, it would take over 73,000 years to reach it!

Where are the Twin Voyagers Now?

On June 2010, Voyager 1 was about 113.731 AU (a bit over 16 billion km, 10.52×10^9 miles, or 0.0018 light-year) from the Sun, the farthest from Earth as compared with any other robot sent so far. Voyager 2 was at a distance of 92.394 AU.

Voyager 1 and Voyager 2 already passed through the heliosphere but were moving in different directions. Voyager 1 escaped the Solar System at a speed of about 3.6 AU per year, while Voyager 2 escaped at a speed of about 3.4 AU per year.

Updated data at www.heavens-above.com

Robotic exploration of the Moon was restarted in the 1990s with spacecraft from the U.S., Japan, and the European Union, providing a wealth of the data, including the discovery of water ice in the craters at the lunar poles, the region that remains hidden from sunlight.

Exploration of Mars has been equally exciting. Americans, Russians, and Europeans sent various space robots and, although some landing missions failed, others were highly successful. For example, the U.S. Viking spacecraft (1975-1983) soft-landed and provided most of the database about Mars collected through the early 2000s. The U.S. Exploration Rovers, known as Spirit and Opportunity, arrived to the Red Planet in 2004 and after more than five years are still active.

Another extraordinary and successful space exploration mission to Mars was NASA's Phoenix Mars Lander, which surpassed its original three-month mission, working for additional weeks in the Martian northern plains. The 350-kg lander was instrumented to study the frozen terrain and habitability potential of the Martian rich soil. Phoenix measures about 5.5 m (18 ft) long with the solar panels deployed and is about 2.2 m (7.2 ft) tall. Launched in August 2007, Phoenix landed on the arctic region of Mars on May 25, 2008, using an improved descent system that combined parachutes and rocket engines.

Shortly after settling on the red surface, Phoenix began transmitting detailed images of the Martian soil, and testing the frozen soil for traces of water. Phoenix completed its activity in August 2008, a few days before its mission was declared officially over, and made a last brief communication with Earth on November 2, 2009 as available solar power dropped with the Martian winter.

Just like Pioneer and Voyager, Phoenix carried "Visions of Mars," a multimedia DVD with works of literature about Mars, and a quarter of a million names of people from all parts of the globe. The DVD was designed to withstand the Martian environment, and is expected to last for hundreds of years.

The Phoenix Mars lander was declared officially dead on May 24, 2010, after repeated failed attempts by the science team to regain contact. A recent image by the Mars Reconnaissance Orbiter showed ice had collapsed one of the lander's solar panels. This turn of events was not surprising, as researchers did not expect it to survive the Martian winter. But Phoenix will be there, waiting until future human explorers go to Mars and find it.

Exploration of other planets continues as well. Early in 2006, NASA's New Horizons spacecraft was launched to explore Pluto and its largest moon Charon; the space probe is expected to approach that remote region of the Solar System by 2015. During its interplanetary cruise, New Horizons already went past Jupiter and provided a bird's-eye view of the giant planet. With a combination of advanced technology, more precise trajectory, and better timing, New Horizons was able to capture more details of Jupiter such as lightning near the poles, the life cycle of fresh ammonia clouds, intriguing boulder-size clumps speeding through the planet's faint rings, and the structure inside volcanic eruptions on its moon Io.

New Horizons continued its interplanetary voyage and by December 2009, it was moving away through interplanetary space between the giant planets in the Solar System. New Horizons is expected to pass the orbit of Uranus on March 18, 2011, and that of Neptune on August 24, 2014. A year later, New Horizons will not land but rather it will approach Pluto, and around July 14, 2015 the spacecraft will conduct its exploration of Pluto and Charon. After nine days of taking data, the robot will leave that zone, moving quietly toward interstellar space, just like Voyager and Pioneer.

In September 2007, the space probe Dawn took off to investigate Vesta and Ceres, two of the largest asteroids or proto-planets in the main asteroid belt, and will travel for an average distance from the Earth of 260 million kilometers. Dawn made its closest approach to Mars on February 17, 2009, getting 549 km from the Red Planet to get a gravity assist boost on its way to rendezvous with Vesta.

Dawn's mission is quite unprecedented, as it will provide data about two of the biggest and more puzzling asteroids in the Solar System. Ceres is so big (with a diameter of about 950 km or 530 miles) that astronomers consider it a dwarf planet. Most intriguing is that Vesta (mean diameter of about 530 km) and Ceres are very different from one another even though they inhabit the same region of the Solar System. Vesta is more like the rocky planets closest to the Sun, and Ceres resembles the icy moons of the outer Solar System. Recent observations revealed that Ceres is spherical, unlike the irregular shapes of smaller asteroids with lower gravity. The instruments that Dawn carries will collect data and images to help us understand why.

Solar sails (also known as light sails or photon sails) will be added to the fleet of space probes in the next few years. Solar sails are spacecraft that use the radiation pressure of light from the Sun to propel the vehicle to high speeds. A solar sail, simply put, is a spacecraft propelled by sunlight. And so it requires enormous reflective surfaces which will be subjected to a steady flood of photons from the Sun. These photons will reflect off the shiny surfaces and impel the spacecraft forward, away from the Sun. By changing the angle of the sail relative the Sun it will be possible to affect the direction in which the vehicle is propelled—in the same manner a sailboat changes the angle of its sails to affect its course.

Japan has led solar sail research and development in recent years and is planning to launch an interplanetary solar sail mission called Ikaros in May 2010. The American Planetary Society is also planning to launch three separate solar sail spacecraft over the course of several years, under a project better known as LightSail.

A different category of space robots that has revolutionized the exploration of space is the orbiting telescopes. NASA's Great Observatories for example, together with telescopes on the ground and automated spacecraft have changed our perspective of the Universe. The Great Observatories are a series of four space-borne observatories designed to conduct astronomical studies over many different wavelengths (visible, gamma rays, X-rays, and infrared). They overlap the operation phases of the missions to enable astronomers to make contemporaneous observations of an object at different spectral wavelengths. The four observatories are the Hubble Space Telescope, the Compton Gamma Ray Observatory, the Chandra X-ray Observatory, and the Spitzer Space Telescope. With the exception of the Compton (which was safely deorbited on June 4, 2000), all other space observatories are still in operation today.

The Hubble Space Telescope, perhaps the better known observatory in orbit, was deployed by a NASA Space Shuttle in 1990. From an altitude of 569 km over the surface of the Earth, the Hubble has made celestial observations and detailed measurements of unprecedented scientific value, revealing the breathtaking beauty hidden in the depths of the cosmos. The Hubble telescope can access the otherwise invisible ultraviolet part of the spectrum, and also has access to areas of the infrared not visible from the ground. Moving around the Earth at 28,163.52 km/h (17,500 miles per hour), the Hubble has made more than 930,000 observations and snapped over 570,000 astonishing gorgeous images of 30,000 celestial objects in its twenty years of operation.

Chandra X-ray Observatory is a telescope specially designed to detect X-ray emission from very hot regions of the Universe such as exploded stars, clusters of galaxies, and matter around black holes. Because X-rays are absorbed by Earth's atmosphere, Chandra must orbit above it, up to an altitude of 139,000 km (86,500 miles) in space. Chandra orbits up to 200 times higher above Earth than the Hubble—about a third of the distance to the Moon! Interestingly, Chandra was deployed on the first shuttle mission commanded by a woman, Col. Eileen Collins, in July 1999.

The Spitzer Space Telescope is an infrared observatory, placed in orbit by a Delta launch vehicle in August 2003. Spitzer maintains a heliocentric orbit similar to Earth's but that makes it drift away from our planet at a speed of 15 million kilometers per year. Spitzer, equipped with a reflector telescope of 85 cm diameter, has a useful life limited by the rate of evaporation of the liquid helium that it uses for cooling. Initially it was expected that the helium would last a minimum of two and a half years and a maximum of five. Yet, Spitzer continues in operation today (summer 2010) being cooled without refrigerant.

The future of astronomical observation is the James Webb Space Telescope, the successor of the Hubble. The Webb will be the size of a tennis court, and it will move in an orbit beyond the Moon. Webb will detect the infrared radiation and will be able to see in that wavelength and in visible light just like Hubble. The infrared vision is essential for our comprehension of the Universe. The most distant objects that we can detect are seeing better in infrared light.

Like its predecessor, Webb will unleash a flood of new discoveries, opening a wide window to a part of the cosmos that is just taking form, making astronomical observations of things not yet observed. This space telescope will travel folded like a cocoon inside a launch vehicle and it will be deployed like a butterfly opening its wings upon reaching its orbit. In 2014 the Webb telescope will be initiated into space, navigating in a distant and lonely orbit where it will look farther.

Robotic spacecraft have explored the planets, moved through inhospitable environments, and other spacecraft are on their way to the stars. Yet, there is so much more to explore, a wide open space with planets, moons, asteroids and space phenomena that only space robots can survey for us, as beyond Mars those are regions to difficult for human beings to reach using current propulsion technology.

Speed Records in Space

The highest speed at which any spacecraft has ever escaped from the Earth is 16 km/s (57,600 km/h or 35,800 mph). This was achieved by the New Horizons space probe launched in January 2006 and now is heading toward Pluto.

The Ulysses probe attained Earth escape speed of 15.38 km/s (55,400 km/h or 34,450 mph). Ulysses is a joint NASA/ESA mission to study the Sun and its influence on surrounding space.

Pioneer 10 attained Earth escape speed of 14.47 km/s (52,100 km/h or 32,370 mph). In its encounters with Jupiter, Pioneer gained additional energy from gravity-assist, enough to escape the Sun's gravitational pull and leave the Solar System.

Dawn broke the record of velocity change by attaining an accumulated acceleration that surpassed 4.3 km/s.

The speed of escape is the initial minimum speed that an object needs to escape the grip of the gravitational force exerted by an astronomical body and to continue moving without having to produce another propulsive effort. In general, the speed of escape is given in terms of the launching velocity without considering aerodynamic friction.

The objects that move at speeds below 0.71 times the speed of escape cannot reach a stable orbit. At a speed equal to 0.71 times the speed of escape, the orbit is circular, and at a greater speed, the orbit becomes an ellipse until it reaches the speed of escape and then the orbit becomes a parabola. For this reason, the speed of escape is also known as parabolic speed.

10
Blazing Trails in the Sky

Fortunate those from Earth that soar in the sky to unravel the mysteries of the heavens

Artist's impression of Orion spacecraft on approach to ISS. Image Credit: NASA.

Our desire to take to the sky must come from observing birds flying. But our yearning to explore the stars that twinkle brightly in the night sky must originate somewhere in a place deep inside our souls. Why else did the ancient peoples of the world weave in their legends and myths tales of people living on the Moon and gods voyaging among the stars? Why else did our ancestors integrate in their concepts of cosmology a belief that our origin is centered in the heavens?

From the beginning of our history, human beings have raised their eyes skywards, awestruck by the exquisite and mysterious beauty of the stars, trying to make sense of what they could see but could not touch. Spaceflight must have been a dream of many, and through the ages it became the fuel that ignited ideas to make it a reality. In the twentieth century, at last, the first people soared high, strapped to thundering machines that left the confines of Earth's atmosphere, reached outer space, and a few lucky astronauts made it to the Moon.

The familiar fable of Daedalus and Iccarus in Greek mythology sparked the original notions of human spaceflight. Through the centuries other thrilling tales took people's imagination much farther. From the extraterrestrial yarn of Lucian of Samosata (A.D. c.120-180), who wrote the first science fiction stories and entertained the notion of humans traveling to the Moon, to Jules Verne (1828-1905) who wrote his action-packed space sagas, all the imaginative ideas of fiction writers made some people wonder if going to the Moon was indeed possible. Of course Verne lived during a time when the scientific theories that led to spaceflight had been cemented by the scholars of the previous centuries. All it took then was a huge imagination and an understanding of the main principles.

The scientific foundation of spaceflight began to take shape back in the sixteenth century when modern astronomy was born. First, Polish astronomer Nicholas Copernicus (1473-1543) determined that the Sun *is* the center of our planetary system. Copernicus heliocentric model, in which the Earth and the planets move around the Sun, began an unprecedented scientific revolution. Scholars continued asking more probing questions, including German astronomer Johannes Kepler who wondered if the Sun exerted a force on the planets and caused them to move. That was before the telescope was invented, and so Kepler was only aware of five planets besides the Earth—Mercury, Venus, Mars, Jupiter, and Saturn—which were known since ancient times.

Kepler discovered that the planetary orbits were not circles, as scholars had believed since the time of Ptolemy. Kepler determined that the paths followed by the planets were actually ellipses and that the Sun was at a focus of the elliptical orbits. This became Kepler's first law of planetary motion. Kepler also discovered the second law, establishing that the velocity of a planet in its orbit increased or decreased while moving closer or farther from the Sun. Finally, Kepler discovered the third law, the one that governs how the velocities of the planets are related. And so it's amazing that, without knowing about the existence of planets beyond Saturn, Kepler's laws described the entire Solar System and paved the way for Newton's work.

Isaac Newton discovered that all motion in the Universe obeys three principal laws. He also found that the force the Sun exerts on the planets is the same force of gravity that keeps us firmly on the ground. In his monumental book the *Principia,* published in 1687, Newton presented a mathematical model of the world in which he stated the physical laws that govern the motion of all bodies on Earth, and the motion of the distant planets.

Newton invented the calculus and discovered the law of universal gravitation. He described gravity with this universal law and told us that gravity is a force that should behave in similar ways regardless of where we are. Newton realized that the gravitational force accounts for falling bodies on Earth as well as the motion of the Moon and the planets in orbit. This was a revolutionary conclusion, as it extended the influence of earthly behavior to the realm of the heavens.

With the three laws of motion, Newton laid the ground for classical mechanics. These principles describe the motion of macroscopic objects, the basis for modern engineering. Newton built much of his knowledge on the ideas of the great Italian scholar Galileo Galilei (1564-1642), who died the same year Newton was born. In formulating his physical theories, Newton set up the foundation of spaceflight and described the operating principle of space rockets.

Following Newton, many other scientists developed the mathematics and scientific principles to grow the branches of physics that were combined to establish the foundation of astronautics and culminated in Einstein's theories of relativity, a new theory of gravity, and his profound insight that gave us a clearer view of the cosmos. The turn of the twentieth century ushered a new revolutionary era for humankind, one that would take people to the Moon and would give everyone on Earth a closer look of the stars.

Pioneer Rocket Scientists

At the same time that the Wright brothers achieved the first powered flight through the air, a Russian schoolteacher named

Konstantin Tsiolkovsky (1857-1935) stunned his compatriots with the idea of spaceflight. In 1905, Tsiolkovsky published an article entitled "*The Exploration of Cosmic Space by Means of Reaction Devices*" in which he envisioned the possibility of space travel and considered the rocket as a means of propelling a spacecraft. He developed the basic equation for reaching space by means of a chemical rocket. Tsiolkovsky calculated the speed required for a minimal orbit around the Earth and predicted that it could be achieved with a multistage rocket fueled by liquid propellants. However, Tsiolkovsky never built or experimented with rockets. The first rockets for space exploration were actually built years later by America's first rocket scientist, Robert Goddard (1882-1945).

In 1912, Goddard proposed the idea of using a chemical rocket to attain Earth's escape velocity. He began to experiment with solid propellant rockets, and in 1919, he published "*A Method of Reaching Extreme Altitude*" in which he explained how scientific instruments could be sent into the stratosphere using a rocket-powered vehicle. In that paper, Goddard described his theories of rocket flight, his research to develop solid and liquid propellant rockets, and the possibilities he saw of exploring space. Goddard developed these ideas independently, and in 1926 he built the world's first liquid propellant rocket. From 1926 and until his death in 1945, Goddard designed, built, and launched many chemical rockets that attained speeds of up to 885 km per hour (550 mph), a world record at that time.

Meanwhile in Germany, Hermann Oberth (1894-1989) was thinking along the same lines. In 1923, when his doctoral dissertation on spaceflight was rejected at the University of Heidelberg, Oberth published a 92-page article entitled "The Rocket into Interplanetary Space" (*Die Rakete zu den Planetenraumen*). Six years later, he wrote a book on the same topic. Oberth's work undoubtedly inspired many of his compatriots, some of which worked for the German army at Kummersdorf near Berlin. Among those rocket scientists was Werner von Braun (1912-1977), one of the developers of the first American space launchers. In 1937, a group led by von Braun began research to perfect the liquid-propellant rocket. Eventually they built the V-2, the world's first long range ballistic missile, launched in 1946 for the first time at White Sands, and finally was used with devastating consequences during the end of World War II.

France also had its spaceflight pioneer. Robert Esnault-Pelterie (1881–1957) was an engineer, aviator, aircraft designer, and spaceflight theorist. He became interested in space exploration and studied rocket flight. On June 8, 1927, Esnault-Pelterie gave a talk before members of the French Astronautics Society on "The Exploration of the Very High Atmosphere by Rockets and the Possibility of Interplanetary Travel" (*L'exploration par fusées de la très haute atmosphère et la possibilité des voyages interplanétaires*). In this talk, published a year later, Esnault-Pelterie considered the exploration of outer space using chemical rocket propulsion. He also discussed technical issues related to the thrust, and guidance and control of a vehicle that could be built for interplanetary travel. Esnault-Pelterie also coined the term "astronautics," which means—literally—navigating among the stars.

People in other parts of the world were also fascinated by the possibilities of space exploration. Thus, it is not surprising to discover a rocket pioneer in Latin America. Between 1946 and 1970, Ricardo Dyrgalla (1910-1970), a developer of liquid and solid propellant rockets, helped Argentina and Brazil to establish their own rocket programs. Ricardo Dyrgalla, whose birth name was Ryszard Dyrgalla, was born in Poland in 1910, but immigrated to Argentina in 1946 upon accepting a job offer from the Argentinean Army. Dyrgalla proposed the development of an aerial-launched "aeromóvil", with a liquid rocket engine totally developed in Argentina. This new vehicle was called AN-1 Tábano (named in honor of the founder of the Argentine Interplanetary Society, Teófilo Tabanera). The Tábano was built and flight tested in 1950, with a glider prototype and later with the rocket engine installed. Dyrgalla also developed the Prosón, a solid-propellant rocket for meteorological research.

Ricardo Dyrgalla wrote several books in the field and was member of the only South American space society at that time, the Sociedad Argentina Interplanetaria (SAI), founded by Teófilo Tabanera in 1951. Tabanera became one of the founding members of today's International Astronautical Federation (IAF). The SAI was instrumental in the development of space sciences and technology in Argentina and was the proponent for the creation of the national space agency, the Comisión Nacional de Investigaciones Espaciales (CNIE) founded in 1960.

Spaceflight owes its progress and spectacular achievements to the efforts of many individuals who discovered the scientific principles, developed the mathematics, the engineering tools, and the technology to make it possible. Building upon the foundation of the scholars of long ago, the rocket scientists, mathematicians, scientists, physicists, and engineers of the twentieth century they all combined their work and vision to transform our world.

Research and development of chemical rockets continued after the Second World War, with the former Soviet Union and the United States developing the first Inter Continental Ballistic Missiles (ICBMs). Engineers in both nations realized that the rocket motors developed for the ICBMs were powerful enough to be considered for space flight.

The Space Age was born with the launch of the first Sputnik in 1957. From then on, the technology of rocket propulsion for space exploration developed rapidly. Powerful solid and liquid propellant rocket engines were designed and built, including the giant Saturn V rocket that sent the first Americans into orbit.

The first men ever to reach space were Russian cosmonaut Yuri Gagarin and American astronaut Alan Shepard. Gargarin was launched into orbit on April 12, 1961 aboard Vostok 1 and circled the Earth for 108 minutes. Shortly after, Shepard performed a 15-minute sub-orbital flight on May 5, 1961. Two years later, the first woman in space was Russian Valentina Tereshkova.

The first Moon visitors were American astronauts Neil Armstrong and Edwin Aldrin. They landed on the Moon in 1969 as part of the historical American mission Apollo 11. Those courageous explorers hopped on the powdery surface of the Moon and collected soil and rock samples that were brought home to study.

The last lunar landing occurred in 1972 during the Apollo 17 mission. It was the eleventh manned space mission in the NASA Apollo program, and the sixth and final lunar landing. During the American Apollo program, twenty-four astronauts left Earth's orbit and flew around the Moon (Apollo 7 and Apollo 9 just made it to low Earth orbit). Twelve of those astronauts landed on the Moon and walked on its surface, and six of those drove a lunar rover. From then on, human missions have been limited to trips to Earth orbit. And finally, twenty two years after the first man reached space, Sally Ride became the first American woman astronaut. After Ride's historic flight aboard the Space Shuttle Challenger in 1983, many other women joined the ranks of space explorers.

Since the Apollo days, the United States and the Soviet Union have been leading the space race. In 1975, the establishment of the European Space Agency added its scientific and technological resources to the space superpowers to pursue the exploration of space. In 2003, the People's Republic of China embarked in the adventure of human space exploration when they launched their first crewed spacecraft named Shenzhou. A second successful piloted mission was carried out in 2005, and a third in 2008 with a crew of three. Chinese astronauts are known as *yuhangyuans* or *taikonauts*. The word *yuhangyuan* means space navigator, while the word *taikonaut* is derived from *taikong*, the Chinese word for space.

During the three-day mission, the three taikonauts launched a small satellite and conducted their country's first space walk. China's space exploration program includes human missions to the Moon and the asteroids, and a robot mission to Mars in the next few years. China is now developing a space station called Tiangong 1.

Several other countries, including Japan, India, Iran, Malaysia, and Turkey, have begun human spaceflight programs and are developing their own launch capabilities. In Latin America several countries have also established space exploration plans. México, Costa Rica, and Argentina have national space agencies to carry out activities of space exploration. In the next decades the number of nations capable of developing a space industry will increase, and the exploration of space will accelerate. Hopefully, this will increase more cooperation and ease political conflicts among countries.

Exploration of space in the future must become more of an international endeavor. The world has witnessed how the Russians and Americans began first as competitors but now they train together and work together to develop space technology and to carry out space missions. Of course there is still work to do to facilitate more international cooperation, but I believe most people in every nation is open and willing to collaborate and join their resources and expertise to explore together the boundless sky.

Spaceflight

Spaceflight is considered when a vehicle travels beyond the Earth's atmosphere. It can be as short as access to Low Earth Orbit (LEO), a three-day trip to the Moon or as long as a voyage across the planets and beyond. Spaceflight involves a journey either by automated or robot spacecraft or voyages by people.

We define LEO as an orbit within the locus extending from the ground up to an altitude of 2000 km. Given the rapid orbital decay of objects below 200 km, the commonly accepted definition for LEO is between 160 and 2000 km (100–1240 miles) above the Earth's surface. The International Space Station moves in a LEO that varies from 319.6 km to 346.9 km above the ground. The Hubble Space Telescope orbits higher—569 km (353 miles) above the surface of the Earth.

Traveling in outer space is a rather complex endeavor. Interplanetary space is dominated by the gravitational fields of the Sun and the planets. The Earth and all other objects in space move in unseen orbital paths and are subject to gravitational forces. Thus, for interplanetary flight we must design a trajectory linking two bodies that move on separate elliptical paths around the Sun.

The first stage of spaceflight is the launch of the spacecraft. To overcome the grip of Earth's gravity, we need a powerful propulsion system producing the required thrust force to ensure the vehicle lifts off the ground and reaches orbit. If the rocket engine is too slow, then the vehicle cannot escape the gravitational field and falls back to the ground. On the other hand, if the rocket departs with sufficiently high speed, the vehicle can escape the gravity of the planet and continue its travel forever, unless it is slowed down by its own propulsion, or by the effect of gravity of a large body (such as a planet or moon) on its path.

The vehicle must be launched at the right velocity, which we call the "escape velocity." Escape velocity, V_s, is the outward velocity required for the vehicle to leave the surface of a planet or moon of mass M and radius R, to escape to infinity (not fall back). The formula for the escape velocity, derived from Newton's law of universal gravitation, is simply $V_s = (2GM/R)^{1/2}$, where G is the constant of universal gravitational. For example, trips to LEO require an Earth escape velocity of 34,920 km/h (9.7 km/s). Compare this with the velocity of a commercial jet airplane, which can reach a cruise speed of just 900 km/h within the atmosphere. Quite a difference, don't you think?

A spacecraft going on a mission from the Earth to the Moon needs an escape velocity of about 40,320 km/h (11.2 km/s) to leave Earth's field. The return flight requires less rocket power, as the escape velocity from the Moon is just 8,640 km/h (2.4 km/s). This is because the Moon is smaller (about a quarter the diameter) and less massive (about one percent the mass) than the Earth.

The flight performance of rocket vehicles is an essential and integral part of rocket science. The flight performance includes establishment of the forces that cause rockets to go into motion along a certain specified flight plan or trajectory. One fundamental force is the thrust produced by the rocket. All the forces are used to derive the equations of motion and the mathematical solution of these equations provide the performance parameters that define a particular flight trajectory. In turn, these parameters describe and represent the flight performance. Let's take a brief view of the basic equation of motion.

Sir Isaac Newton and Rocket Science

We apply Newton's laws of motion to design rockets and to derive the equations of motion that govern spaceflight. These three laws are fundamental statements of how things move; they apply to rockets, launch vehicles, and all spacecraft. Newton's laws are independent of the type of energy used in the rocket. However, since we use chemical rockets to launch a large vehicle to accelerate it to the required orbital speeds or to leave the gravity field of the Earth, let's begin with an overview of chemical rockets and illustrate the application of these physical laws.

A chemical rocket is a heat engine where the propellant—a chemical mixture of fuel and oxidizer—burns in a combustion chamber at high pressure. To produce thrust (the force that propels a spacecraft), the hot combustion gases exhaust through a nozzle at the rear of the rocket, accelerated at high velocity. The result of the acceleration of the hot gases is a force (thrust) exerted by the rocket in a direction opposite to that in which the gases move out.

The chemical propellant can be in liquid or solid form, and so we design solid rocket motors that burn solid propellants, and liquid rocket engines that operate with liquid propellants. Typical space launchers use a combination of solid and liquid rockets to boost a spacecraft to orbit. For example, the Space Shuttle Orbiter is launched vertically boosted by two solid propellant motors, known as the Solid Rocket Boosters (SRBs), and powered by three liquid propellant rockets known as the Space Shuttle Main Engines (SSMEs) that burn liquid hydrogen and liquid oxygen.

Now, let's interpret Newton's laws of motion. The first law, known as the law of inertia, tells us that the propulsion system of a launch vehicle must develop enough thrust force to overcome the force of gravitational attraction of the Earth. The rocket engines must start the vehicle moving and accelerate it to the required velocity.

Newton's second law of motion is the momentum law, which states that when a force is applied to a body, the time rate of change of momentum is proportional to, and in the direction of the applied force. In Newton's laws of motion one basic concept is force, the cause of movement; another concept is mass, the measure of the amount of matter that is put in motion; the two are denominated with the letters F and m, respectively, and are related by the acceleration, a, in the second law equation, $F = m \cdot a$, which is the simplified relation we all learn in school. A better mathematical form is $F = \frac{d(mv)}{dt}$, as it represents the time rate of change of momentum. This is one of the fundamental equations in rocket science from which we determine the thrust force required to launch a vehicle.

As we said, a rocket-propelled spacecraft derives its acceleration from the exhaust of gases produced by the burned propellant and its ideal equation of motion (neglecting gravitation and drag forces) follows directly from the conservation of the total momentum of the vehicle and the exhaust gases:

$$m\frac{dv}{dt} = \frac{dm}{dt}V_{ex}$$

where m = time varying (instantaneous) mass of vehicle
dv/dt = vehicle acceleration
dm/dt = rate of change of spacecraft mass due to propellant burnt
V_{ex} = velocity of exhaust gases leaving the rocket

The right hand side of the above equation (product of the rate of mass exhausted and the exhaust velocity) is the thrust force generated by the rocket propulsion system:

$$F = \frac{dm}{dt}V_{ex}$$

The thrust can be treated, for most purposes, as if it were an external force applied to the vehicle. In its idealized form, $F = \dot{m}V_{ex}$ represents the thrust force of a rocket, and $\dot{m}$ represent the exhaust flow rate (assumed in this case to be constant during the thrusting time) of the hot gases expanded and expelled through the nozzle. Thrust is a total force in a particular direction. Therefore, the units of thrust are the same as those of force: newtons (N) in the SI system and pound-force (lbf) in the English system.

Finally, Newton's third law of motion, known as the action-reaction law, explains the principle of operation of all propulsion systems, as they propel a vehicle in one direction by the force of expelled matter in the opposite direction. In fact, chemical rocket engines develop thrust by expelling the burnt propellant gases at high velocity through an exhaust nozzle. The reaction force (thrust) acts in the opposite direction of the exhaust hot gases.

For a chemical rocket, the thrust depends on how fast the hot gases leave through the nozzle, and thus the design of this part of the propulsion system is crucial to help us optimize the performance of a chemical rocket engine. A rocket nozzle is a converging-diverging duct that connects to the combustion chamber and through which the hot gases expand and accelerate to the required exhaust velocity V_{ex}. Although the energy produced by the chemical reaction of propellants varies to a certain extent, the maximum exhaust velocity that can be delivered in modern rocket engines is 3200 m/s with solid propellant, and up to 4500 m/s with liquid propellant.

The ratio of the area of the narrowest part of the nozzle (throat) to the exit plane area determines—to a great extent—how efficiently the expansion of the exhaust gases is converted into the linear velocity *V*. The second law of motion tells us that there is an exchange of momentum between the rocket engine and the exhaust mass of gases (see the previous equation). However, as the propellant is consumed, the rocket will accelerate faster, and the acceleration of the vehicle will increase. We take this in consideration when we calculate the final speed of the vehicle after some or all of the propellant has been consumed.

An effective nozzle has a very large area ratio to maximize thrust. See the picture below to get a sense of the size of the nozzle for the Space Shuttle Main Engine (SSME), which has an expansion area ratio of 77.5. Furthermore, rocket engines must produce high thrust to compensate for the overall weight of the spacecraft. The total weight of the vehicle includes the rockets, the propellant and the storage tanks, the structural mass, and the payload. Chemical rockets have a relatively high thrust-to-weight ratio (F/W = 50 – 75), a desired characteristic for a launch vehicle to ensure it will overcome Earth's force of gravity and accelerate into orbit.

Space Shuttle Main Engine (SSME) Nozzle. Credit: NASA

The rocket concept as described above appears simple at first glance. However, the design, analysis, and manufacture of a practical and effective rocket engine form part of an enormously complex engineering process. It requires the expertise of many engineers and scientists, and invests many years of technology development to go from a concept on paper to an efficient and reliable operational propulsion system that can deliver a payload to orbit or move a vehicle in a given space trajectory. Since most of us are familiar with NASA's Space Shuttle Orbiter (the spacecraft that carries the payload and transports the astronauts), we use it to give an idea of the enormous forces required to ascend from the ground to Earth orbit.

The Space Shuttle, carrying a payload of up to 29,400 kg, has to accelerate by 25,880 km/h (7.18 km/s) to reach its orbital speed, a speed which ranges from 27,355 km/h to 28,960 km/h (7.6 – 8 km/s). To reach LEO, the propulsion system must deliver more than 35,585 kN (or 8 million lbf) of thrust. The three Space Shuttle Main Engines (SSMEs) at the rear of the Orbiter fire together, producing 31,000 kN (7,000,000 lbf) of thrust. Each solid rocket booster has a liftoff thrust of approximately 12.5×10^6 N (2,800,000 pounds-force) at sea level, increasing shortly after liftoff to about 13.8×10^6 N (3,100,000 lbf). The two SRBs, with their combined thrust of 26,000 kN provide most of the power (more than 70 percent of required thrust) for the first 2 minutes of flight. The remaining 30 percent thrust is provided by the liquid propellant SSMEs. The two SRBs take the Space Shuttle Orbiter to an altitude of 45 km.

Spaceflight Performance

The flight performance of a rocket vehicle is interrelated to the engine performance, and in many cases the parameters which appear in the flight plan are related to the rocket, as for example the exhaust velocity. The flight requirements of a rocket which is to perform a specified mission (i.e., to travel from a starting point and reach another) must in some way dictate the requirements of thrust.

The rate at which a spacecraft moves in space is very important, as the distances to be traveled are extremely large. We can derive a mathematical relation that tells us how fast a vehicle can move. Starting with the thrust equation, we perform an analysis (using the methods of calculus) and obtain the rocket equation (assuming exhaust velocity is constant during the thrusting time): $\Delta V = V_{ex} ln\left(\frac{m_o}{m_f}\right)$.

This relation, known as delta-V, implies that the spacecraft experiences an increment in velocity which is linearly dependent on the exhaust velocity and logarithmically dependent on the mass exhausted through the nozzle. Although idealized, the rocket equation helps us illustrate how the incremental velocity required during the powered segment of a flight mission is mainly a function of the exhaust velocity V_{ex} of the gases coming out the engine, and of the mass ratio $\frac{m_o}{m_f}$, where m_o is the initial or loaded vehicle mass (engine + propellant + payload), and m_f is the final or empty mass of the vehicle, also known as mass at cut-off (engine + payload).

We can also write the rocket equation as $\frac{m_o}{m_f} = e^{\Delta V / V_{ex}}$. This relation neglects effects of gravity, drag and other losses, and it assumes a constant exhaust velocity relative to the rocket. However, it's an acceptable first step to determine a great deal about the performance of a rocket propulsion system for a given space mission.

The ΔV is an equivalent velocity change that depends on the destination, travel time, trajectory, gravitational force field, and the thrust mode for a given space mission. A higher exhaust velocity is always desired, because less propellant will be required for a given mission. If a spacecraft starts initially from rest, then the speed of a vehicle at burnout (when propellant is exhausted) is $V = V_{ex} ln\left(\frac{m_o}{m_f}\right)$. For example, if a vehicle with a mass ratio equal to 20 is launched from Earth using a rocket that expands the hot gases at 3 km/s, it can attain a velocity of 8.98 km/s at the end of the burn. In this case, delta-*V* is the same as the escape velocity.

Delta-*V* is a scalar measure for the amount of "effort" needed to carry out an orbital maneuver, i.e., to change from one orbit to another. The time-rate of change of delta-*V* is the magnitude of the acceleration caused by the engines, i.e., the thrust per kilogram of total mass. Electric propulsion is characterized by its high delta-*V*, and so it is commonly used to power space probes and satellites. Chemical propulsion, on the other hand, is characterized by high thrust, and that is why chemical rockets are used for launching large spacecraft.

When designing a trajectory for a spacecraft, we use delta-*V* as an indicator of how much propellant will be required. Actual propellant usage depends linearly on delta-*V* because not only the rocket but also the propellant in the tanks has to be accelerated. As the propellant burns and is exhausted through the rocket nozzle, the mass of the rocket reduces and, therefore, achieving the necessary delta-*V* for a maneuver will require less propellant.

We can increase the mass ratio by increasing the propellant to overall mass ratio. This can be accomplished by either increasing the mass of the propellant without increasing the size of the vehicle, or by decreasing the overall initial mass by the use of lightweight structures and other engineering or design improvements.

In general, V_{ex} is given in terms of the specific impulse I_{sp}, a very important parameter used to compare the performance of rocket engines. The specific impulse is properly defined as the impulse (I) given to the rocket per unit weight of propellant. As the impulse is the effect of a force applied for a very short time, $I = F \cdot dt$, and the weight of propellant during that short time is simply $mg \cdot dt$, we derive an expression for the specific impulse: $I_{sp} = \frac{F}{\dot{m} \cdot g} = \frac{\dot{m} \cdot V_{ex}}{\dot{m} \cdot g} = \frac{V_{ex}}{g}$, where g is the acceleration of gravity ($g = 9.80665$ m/s^2 on the surface of the Earth). The real units of specific impulse are newton-seconds/newton, and cancellation of units produces seconds.

The specific impulse is indispensable in evaluating the capability of a rocket engine as a propulsion system. The specific impulse is also considered as a measure of the fuel efficiency of the rocket. For example, when we say a rocket engine has a specific impulse of 408 seconds we mean to say that it produces 4000 kN of thrust for every kilogram of propellant burned in 1 second.

The computation of specific impulse is not trivial. It requires a complex thermochemical analysis to determine the equilibrium state of a reacting propellant mixture, which depends on pressure, temperature, and mixture ratio. The analysis is rather complicated and requires the use of numerical methods to solve the combustion reactions and, at the same time, solve the fluid dynamic equations that model the fluid flow through the engine. However, to get started we use generalized methods of solution based on dimensionless analysis, and we use extensive empirical experimental data for many propellant mixtures used in the course of developing the potent chemical rockets built to date.

We aim to design a rocket with high I_{sp} in order to minimize the propellant requirements of a space mission— high I_{sp} provides high exhaust velocity V_{ex}. However, because there is a limited energy release in chemical reactions and because a thermodynamic nozzle is used to accelerate the combustion gases, the specific impulse of a chemical rocket is limited. The maximum I_{sp} that can be achieved with liquid propellant rockets is between 400 and 500 seconds. Each SSME has an $I_{sp} = 453$ s (in space) and $I_{sp} = 363$ s (at sea level).

Let us assume we design a rocket engine with a specific impulse of 450 s, and a mission requires a ΔV of 10 km/s (typical for launching a vehicle to LEO), then the mass ratio of the spacecraft would have to be $m_o/m_i = 9.63$. The problem we would face here is that most of the vehicle mass would be propellant, and due to limitations of the strength of materials, it may be impossible to build such a vehicle just to ascend into orbit.

One way to get around this problem is by building a rocket engine in several stages, throwing away the structural mass of the lower stages once the propellant is consumed. The result is higher mass ratios, and hence a space mission can be achieved with low I_{sp} engines. The Space Shuttle uses a multi-stage rocket propulsion system. The first stage consists of the two solid propellant rocket boosters (SRBs) used to increase the overall thrust of the vehicle during lunch. As you probably noticed when observing the launch of the Shuttle, the SRBs are discarded after two minutes; that's when their solid propellant is completely burned out. The SSMEs continue firing for about eight minutes. They shut down just before the spacecraft attains orbital velocity. The external tank then separates from the Orbiter and follows a ballistic trajectory into a remote area of the ocean.

When Apollo 11 took the three astronauts on the first mission to land on the Moon, the Saturn 5 rocket first carried the spacecraft to an altitude of 185 km (LEO), moving a speed of 24,800 km/h (6.88 km/s) —just below orbital velocity. Then the third stage fired briefly, enough to accelerate the spacecraft to the required speed. The engine was shut down, allowing the vehicle to coast in that Earth orbit. After performing the required maneuvers to line up the spacecraft on the correct flight path, the third stage of the rocket was restarted, increasing the speed to about 10.86 km/s (39,100 km/h), the escape velocity at that altitude required to completely overcome the influence of Earth's gravity.

For three days, Apollo 11 coasted around the Earth spiraling up on its way to the Moon, until the pull of our planet's gravity became weaker and the spacecraft slowed down. By the time Apollo 11 was about 346,000 km from Earth its speed had decreased considerably. Then the gravity of the Moon began to pull the spacecraft in, increasing its velocity. Once in lunar orbit, the astronauts separated the lunar module, fired its descent stage, and began the landing maneuver. When the lander was about 1.5 meters from the lunar surface, the engine shut off, and the vehicle carrying two people touched the surface of the Moon.

It sounds easy, but the maneuvers that resulted in such a momentous milestone for humanity relied in mathematically precise calculations, simulations, and extensive planning, deriving the analysis from orbital mechanics and rocket science. As noted earlier, the movement of every object in space is influenced by gravity and follows orbital paths—around planets, moons, and the Sun. Planets move in elliptical orbits with the Sun at one focus. Because the planets are at different distances, another challenge is to travel to a destination that is moving around the Sun at a different speed than the Earth.

The orbital speeds of Moon and Earth must be considered in calculating the time when the spacecraft intersects the orbits. Thus, the launch date has to be carefully selected to ensure that both the Earth and the target destination are aligned in the most favorable position when the spacecraft begins the trip.

Consider this: the launch pad is spinning at about 1670 km/h (1037 mph) at the time when the rocket is launched. The Earth is also moving around the Sun at about 107,300 km/h (67,062 mph). At the same time, the place where astronauts are going (target destination) is also moving with its own orbital speed. Let's say you're going to the Moon, which orbits around the Earth at approximately 3600 km/h (1 km/s) average speed. Your rocket vehicle is moving at 24,800 km/h. Can you see the complication? It's like trying to shoot a basketball while running in circles and the hoop pole is moving at a different speed.

So far, we addressed the thrust required during the launch phase of spaceflight. What about landing on an extraterrestrial body? This is not an easy maneuver either, as spacecraft do not have the landing gear of aircraft designed to touchdown on specially built landing strips or airport runways. Spacecraft attempting to land on an unfamiliar extraterrestre surface need to perform a maneuver called deep throttling. While landing on a planet, moon or an asteroid, the propulsion system has to be able to throttle up quickly and accurately to avoid damaging the craft or the crew.

Deep throttling refers to the thrust capability that enables a spacecraft to maintain adequate thrust for in-space travel, yet be able to power down for a precise, gentle landing on any unfamiliar surface. The propulsion system must have the ability to deliver a thrust range from 104 percent of rated power down to 5.9 percent. This equates to an unprecedented 17.6-to-1 deep-throttling capability requirement. Deep throttling is one of the key propulsion technologies that are being developed for the rocket engines of the future.

Rocket science is complex, and it requires many years of studies to apply the fundamental theories of physics and engineering principles to design rocket concepts, to model them with advanced computational tools, to experiment and develop them to become effective propulsion systems capable of delivering human and material payloads to Earth orbit and beyond.

As a final note I should add that the rocket theory discussed in this chapter is not applicable when spaceflight velocity equals or exceeds roughly ten percent of the speed of light because then relativistic effects must be included. That is the rocket science we will need for the future, when we develop relativistic rockets for interstellar travel.

11
From Earth to Mars

"The Earth is the cradle of humanity, but mankind cannot stay in the cradle forever."
K. Tsiolkovsky (1896)

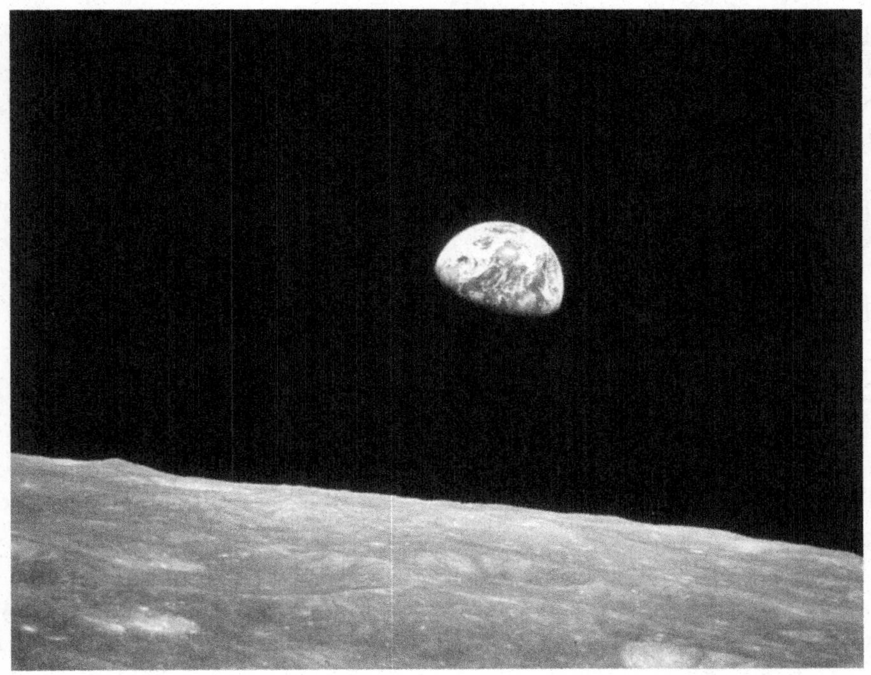

Earth Rise. Photo taken from the Moon. Credit: Apollo 8 Crew, NASA.

The United States has a new strategy for human exploration of space. Instead of going back to the Moon and continue flying the Space Shuttle to transport astronauts to the International Space Station, NASA has received a mandate to cancel the operation of the Space Shuttle, and to revise the options for human spaceflight. In the plans of 2010, NASA included the cooperation of the private industry to develop a space taxi to ferry astronauts to orbit, and it considered new interplanetary missions with travel to the asteroids and ultimately to Mars.

The new NASA plan for human spaceflight focuses on technologies and milestones that will advance piloted space flight out of Earth orbit and into the Solar System. Mars is the ultimate goal, of course, and the path to get there is said to be flexible, to be determined step-by-step. NASA has established some deadlines for developing the new technologies—design the new rockets capable of making the trip to deep space by 2015. NASA has also set destinations with deadlines—land on an asteroid by 2025, and reaching Mars by 2035. Two years ago, NASA was preparing to return to the Moon and build a permanent settlement there. Today, it is not clear whether the Moon is a step toward Mars or a destination by itself.

In the next few pages, I will highlight some components of the new plan for human spaceflight, starting with the launch capability required to ferry the astronauts back and from the International Space Station, after the Space Shuttle is retired, and ending with highlights of the plans for an expedition to the asteroids and a voyage to Mars.

Space Taxis

NASA's Space Shuttle, officially named the Space Transportation System (STS), has been the launch vehicle of the government of the United States since the mid-1970s. During its almost a quarter of century operation, the Space Shuttle not only has facilitated research that has produced countless technologies, many with application in the Earth, but also it has made human spaceflight to Earth orbit almost routine.

According to the 2004 Vision for Space Exploration, the use of the Space Shuttle was to be mainly focused on completing assembly of the Space Station by 2010, after which it was to be retired from service, and eventually would be replaced by the new Orion spacecraft (now slated for cancellation). The Orion Crew Exploration Vehicle (CEV) is a capsule-type crewed spacecraft that is being developed as part of NASA's Constellation Program. In this plan that includes a new two-vehicle launch system, one of which would be used to deploy Orion in its trip to the ISS, and with modifications would be used to make the trip back to the Moon. The crew launch system, known as Ares 1, was planned to be ready for its first piloted space flight in 2015.

On February 1, 2010, the President of the United States announced a proposal to cancel the NASA's Constellation, effective with the U.S. 2011 fiscal year budget but later announced changes to the proposal in a major space policy speech at Kennedy Space Center on April 15, 2010. It is still unclear at this time which among the original components of the proposed two-launch systems will continue to be developed.

The idea of a space taxi arouse from the need to fly astronauts to and from the International Space Station after NASA retires its fleet of space shuttles in 2011. And because there is no ready replacement for the Shuttle, after it is retired NASA plans to buy rides for astronauts on the Soviet Soyuz, until a new American system of rockets and spacecraft or capsules for the crew is ready. The Soyuz spacecraft has been used since 1967 for both long-haul flights into orbit with two or three cosmonauts on board.

Currently, the private industry is engaged with NASA to develop rockets and spacecraft to deliver crew and cargo to the Space Station. Many potential launch vehicles that could replace the Shuttle have been considered. Most of those design concepts represent the traditional conservative approach of making incremental improvements on already existing systems. There are a number of proposals from the private sector, each one offering capability to transport astronauts and cargo payload. The front-runners in the space taxi competition include Delta II and Delta IV rockets, SpaceDev Dream Chaser, and the newly developed two-stage rocket Falcon 9.

The Sierra Nevada Corporation is developing SpaceDev Dream Chaser, a vehicle that is derived from an existing NASA design, which has the objective to reach a suborbital altitude of approximately 160 km (100 miles) and to reach a maximum orbit to 420 kilometers (around 200 miles) over the Earth surface. The Dream Chaser is powered by hybrid rockets, each one able to produce approximately 444,822 N (100,000 lbf) of thrust. The hybrid rocket is a motor that uses propellant in two different physical states—one solid and the other gas or liquid.

Hybrid rockets have some advantages over conventional liquid- and solid-propellant rockets due to their simplicity, safety and cost. As it is almost impossible that the fuel and the oxidant become mixed intimately (being in different states), the hybrid rockets tend to fail more benignly than the rockets with liquid or solid propellants. Just like the liquid rocket engines (and not like the solid rocket motors), the hybrids can extinguish themselves easily and they can also be decelerated. The theoretical specific impulse of the hybrid rockets is generally higher than the specific impulse of solid propellant motors and is equivalent to rockets burning liquid hydrocarbons. It is estimated that hybrid rockets can yield a specific impulse of about 400 seconds with metalized fuels.

A new company sponsored by NASA is Space Exploration Technologies, better known as SpaceX. This company is developing a launch system partially reusable of conventional design known as Falcon 9. The two-stage vehicle utilizes liquid propellants LOX/RP-1, with the first stage propelled by a single Merlin rocket engine and the second stage powered by a Kestrel engine. Propellant RP-1 (Rocket Propellant 1) is a form of highly refined kerosene similar to the fuel used in jet engines. It has a specific impulse lower than liquid hydrogen (LH2), but RP-1 is cheaper, can be stored at room temperature, has less risk of exploding and has a higher density. In volume RP-1 it is more powerful than LH2, and has a fraction of the toxicity and carcinogen risks of hydrazine, another liquid fuel that is used in many rocket engines.

The Falcon 9 is the first launch vehicle that utilizes only liquid propellant rockets (without solid rocket boosters like most other designs) for the entire mission. In June 2010, during a flight test Falcon 9 reached low Earth orbit in nine minutes. This flight demonstrated its capability, putting Falcon 9 at the front of the list of candidates to become the space taxi that could take astronauts to the Space Station. SpaceX is also developing a crew capsule named Dragon, a spacecraft planned to be launched by Falcon under NASA's Commercial Orbital Transportation Services (COTS) program.

Another contender is the Delta IV launch vehicle developed by the Boeing Company. Delta IV can accommodate one or several payloads in the same mission. The rockets can send payloads to polar orbits, solar synchronous orbit, orbits of geosynchronous transference (GTO) and geosynchronous orbit, and of course to low Earth orbit (LEO). The capability of achieving geostationary transfer orbit is critical to the placement of modern satellites, as well as to the success of future space programs going to the Moon, the asteroids, and to Mars.

Each rocket Delta IV is mounted horizontally, is elevated vertically in the launch platform, integrated with its payload, it is fueled and launched. All the Delta IV configurations utilize a common booster core first stage with the Pratt & Whitney Rocketdyne RS-68 main engine, which burns mixtures of LOX/LH2. The second stage has the Pratt & Whitney RL10B-2 upper stage engine and two sizes of fuel and oxidizer tanks

Boeing and Bigelow Aerospace joined forces to develop the Delta II launch vehicles. The Delta II can be configured with two- or three-stages to facilitate a wide variety of mission requirements. In all options, the first stage includes the RS-27A main engine developed by Pratt & Whitney Rocketdyne. For additional impulse during lift-off, the first stage can be configured with three, four, or nine graphite rocket boosters. The Delta II vehicles can carry payloads to LEO, and if it needs to take payload to GEO, then it can use a third rocket stage, which utilizes the Thiokol Star-48B solid rocket motor.

Orbital Sciences Corporation is another company that has its own family of launchers. The Minotaur is a family of solid-propellant rockets derived from the intercontinental ballistic missiles Minuteman and Peacekeeper. The Minotaur can launch small satellites to Earth orbit, including GTO and trajectories that reach the orbit of the Moon.

At the time I wrote this page it was still too early to assess which among the several proposals from the industry would replace the Space Shuttles. However, most of the concepts consider space launchers that look very much like the systems originally proposed by NASA under its Constellation Program, i.e., launch vehicles powered by multi-stage chemical rockets with a capsule-type crew transportation vehicle sitting atop the towering launchers, designs that depart considerably from the shuttle's winged spacecraft approach that allows it to land like an airplane.

In the future, to make spaceflight more economic and to realize space tourism, we need fully reusable space launch vehicles. In order to make it possible, we needed to develop reusable spaceplanes. These are vehicles that take off horizontally from a runway like an airplane, fly to Earth orbit, and return to Earth intact. We have conceived two types of single stage Spaceplanes for LEO transport: (1) Vertical-Take-Off-and-Landing (VTOL) Single-Stage-To-Orbit (SSTO) vehicle, and (2) Single-Stage-To-Orbit (SSTO) Horizontal-Take-Off-and-Landing (HTOL) vehicle. Non-reusable or consumable vehicles SSTO VTOL (vehicles that are used once) are simple to design using existing components. But the re-usable SSTO HTOL vehicles are difficult to develop technologically.

The basic problem for designers of any re-usable spacecraft is to reach sufficient speed to arrive at the required orbit without carrying so much propellant; the vehicle would be too heavy if a large propellant tank is carried along. Therefore, the design requirement is to construct a very light-weight vehicle, or to find a way to carry less propellant. Propulsion designs that combine rockets with jet engines (which ingest the air from the atmosphere) could be used in the first stage of the flight.

The spaceplane requires the development of hybrid, dual mode air breathing/rocket engines. The proposal is to propel a spacecraft at very high speeds through the outer edge of the atmosphere, using air breathing scramjet engines, instead of rockets. A scramjet/rocket-powered spaceplane has the potential to reduce launch costs to LEO. Proposals for such vehicle have been around for over thirty years. Why this architecture is attractive? Consider this: Seventy-five percent of the Space Shuttle's 4.4 million pounds gross lift-off is propellant mass, and of this propellant mass, 83 percent is due to the oxidizer (liquid oxygen or LOX). On the other hand, about twenty percent of the atmosphere is oxygen, and that can be had for free, why not use it?

Thus, a spaceplane powered by an air breathing/rocket hybrid engine can use the atmospheric oxygen to oxidize the fuel while flying within the atmosphere, and then switch to rocket mode propulsion when the air becomes too thin to support combustion in the scramjet. Of course, some on-board oxygen would still be required to feed small rockets used for the final acceleration, on orbit maneuvers, and for re-entry initiation. But carrying a much smaller oxidizer tank would result in enormous weight savings.

At the same time, a spaceplane or SSTO HTOL vehicle would be more flexible to operate, as it could take-off and land from specially equipped airports rather than be restricted to launch sites or ballistically land in the ocean. Also, being an aerodynamic vehicle, a spaceplane would be safer for the passengers, as it would have better abort capability than the Shuttle and other conventional launch systems.

An innovative, reusable space plane called the Skylon is being developed in the United Kingdom, aiming to make payload delivery and passenger space travel easier and more affordable. The key to realizing this goal is a combined cycle engine that can operate both in air-breathing and pure rocket modes. During ascent and flight through the atmosphere, Skylon will get thrust from a unique air-breathing engine called SABRE that will reduce the weight and running cost of the spacecraft by using atmospheric oxygen in the combustion process.

The Sabre engine is essentially a closed cycle rocket engine with an additional pre-cooled turbo-compressor to provide a high pressure air supply to the combustion chamber. The chemical reaction of air and fuel in the combustor in turn produces the hot gases that are expelled at great velocity through the back nozzle to produce the thrust. This allows operation from zero forward speed on the runway up to Mach 5.5 in air-breathing mode during the ascent, increasing the installed specific impulse 3-6 fold. As the air density decreases with altitude the engine eventually switches to a pure rocket, propelling Skylon to orbital velocity. In rocket mode the engine operates as a closed cycle LOX/LH2 high specific impulse rocket engine.

The Skylon spaceplane project is led by a British company called Reaction Engines Limited (REL) in a joint public/private program with funding from the European Space Agency (ESA). The Skylon spaceplane was derived from the British Aerospace HOTOL project with its Rolls Royce RB545 engines. Together the two projects have been under active study for 25 years.

In other spaceplane concepts that use the underside of the vehicle as part of the propulsion system (such as the U.S. National Aero Space Plane NASP of the 1980s), integration is the biggest challenge. The reason why the belly of the vehicle is made to be part of the engine is that, at hypersonic speeds (more than five times the speed of sound), very large airflows are needed to develop the required thrust. Thus, major sections of the vehicle itself must serve as extensions of the engine inlet and nozzle. Also, prior to entering the engine inlet module, the airflow is slowed down and compressed by the shocks produced by the bow of the vehicle.

In the past, attempts to design single stage to orbit rockets have been unsuccessful largely due to the weight of the oxidizer such as liquid oxygen and the required tanks. An obvious solution to reduce the quantity of oxidizer that a vehicle is required to carry is being able to use atmospheric oxygen in the combustion process. That is why I believe Skylon, powered by its innovative SABRE rocket/jet engine, will revolutionize access to space.

Return to the Moon

On July 20, 1969, the first human beings visited the Moon, extending the limit of the human experience 400,000 kilometers into space. The first trip to the Moon is considered a transcendent accomplishment. Five other lunar landings followed between 1969 and 1972. Apollo 17 became the aim of the initiation of the movement of the human race towards the Universe, but unfortunately we did not finish the lesson that would teach us how to inhabit another world in the sky.

We must go back to the Moon, this time to stay and colonize it. We need a sustained human presence in space, and the Moon is the best next place to do that. In 2008, I wrote that by the year 2020, the U.S. and other nations would begin building lunar settlements where people would learn how to live in an extraterrestrial world. I described NASA mission to return to the Moon. Plans for that mission, under the Constellation Program, were abandoned in 2010 and instead NASA is now considering trips to the asteroids and to Mars.

I believe that it is necessary for humans to return to the Moon and build a settlement there in order to learn how to live in another world and also to test the technologies required for the crewed trips to Mars and beyond. Making routine trips to the Moon would also help engineers resolve technical problems and challenges faced by the space explorers during long interplanetary journeys.

Although NASA plans for the exploration of space are changing, I will devote the next few paragraphs to describing the Constellation Program, as the details of the projected mission to the Moon and the conceptual design of the lunar spacecraft and launch systems will serve me to illustrate such endeavor.

With the Constellation Program, NASA intended to make the trips back to the Moon with a crew of four. The astronauts and a lunar module would be launched in a vehicle powered by chemical rockets very similar to those used in the Apollo program. Orion, the new spacecraft designed to carry the four astronauts to and from the Moon was named more precisely the Orion Crew Exploration Vehicle (CEV), while the new lunar lander was named Altair. The CEV and the lunar lander would be launched into orbit by a new launch system which consists of two vehicles named Ares 1 and Ares 5—one to carry the crew, and the second to carry heavy payload. NASA's launch concepts are illustrated in the image below.

A comparison of NASA launch vehicles – Credit: NASA

Going back to the Moon, astronauts could follow a path in space already familiar, travelling onboard a spacecraft similar to the Apollo command module of the first lunar explorers in the 1960s. But there could be some differences as well. Orion is a bit roomier than the Apollo capsule. With a diameter of 5 meters (about 16.5 feet) and a habitable volume of 11 cubic meters (692 cubic feet), Orion was designed to be able to carry four astronauts to the Moon, and also as the Space Shuttle replacement it would transport up to six to LEO.

Orion includes a service module with storage bays to support the main cabin, batteries and fuel cells plus a pair of circular solar arrays to help keep the vehicle powered during its intended long stay in lunar orbit. The UltraFlex disk-shaped solar arrays, measuring about 5.5 meters (18 feet) in diameter, would provide power for Orion during its mission. The solar arrays open like a Chinese fan and are made of ultra-light materials that provide high strength and stiffness, and are very compact when closed; this results in extremely low stowage volume.

On average, the Moon is about 384,400 km (almost a quarter million miles) from the Earth. The actual distance from Earth to Moon varies; sometimes the Moon is closer, and at other times it is farther away. This variation is due to the Moon's elliptical orbit, and it affects the design of the flight path or trajectory the spacecraft must follow after launching from Earth.

A trip to the Moon requires boosting the spacecraft into an Earth-parking orbit, where the spacecraft can coast (no rocket power) until it is in the proper position to begin the second leg of its journey. The Moon is moving around the Earth at an average speed of 3650 km/h (2268 mph). Earth is also revolving on its own axis as well as orbiting the Sun at 106,217 km/h (66,000 mph). At the appropriate time, the spacecraft must accelerate of the parking orbit and into a trajectory or path to its target, the Moon. If you could travel to the Moon on a vehicle moving at a constant speed of 900 km/h, it would take more than sixteen days to get there. On a rocket-powered vehicle, it can take just three days, including launch, orbit transfer maneuvers, cruise, and descent. This time also includes coasting while orbiting the Earth until the spacecraft is in the right position to attain or intersect lunar orbit.

The NASA Mission Plan to return to the Moon in the 21st century can be described with a simplified 2-way flight trajectory. The heavy lifter Ares 5 would be launched carrying the Lunar Lander Altair and the departure stage required to escape Earth orbit. A four-member crew would follow on board the Ares 1 launcher, travelling in the Orion capsule located above the upper stage. While waiting for countdown, the astronauts would be strapped to their seats at a height of about 90 meters over the ground. Once in orbit, the astronauts would connect their capsule Orion with Altair and the departure stage before continuing their journey to the Moon. Three days later, the crew would reach lunar orbit. The astronauts would enter the landing vehicle, leaving behind the capsule in orbit around the Moon.

After descent, astronauts would stay (initially for seven days) on the Moon. After completing the exploration mission, the crew would ascend to lunar orbit, reconnect to the capsule and would return to Earth. After de-orbiting, the service module was designed to expose the thermal barrier for the first time in the mission. The parachutes would be deployed, the thermal barrier would fall, and the capsule would land safely.

The propulsion system for the Ares vehicles consists of solid and liquid propellant rockets. For its first engine stage, Ares 1 will have a single five-segment solid rocket booster, a derivative of the Space Shuttle's SRBs. The second stage (to power the CEV) is a liquid oxygen/liquid hydrogen J-2X rocket engine derived from the J-2 engine used on Apollo's second stage. The Ares 1 vehicle could lift more than 25 metric tons (55,000 pounds) to low Earth orbit (LEO). Since the new spacecraft was to replace the Shuttle Orbiter, Orion was also designed to deliver a six-member crew and supplies to the International Space Station.

Ares 5 is the cargo launch vehicle that would carry the heaviest payload to orbit. The heavy-lifting Ares 5 was also to have two stages. For its initial insertion into Earth orbit, the first stage is designed with a solid rocket booster comprised of two five-and-a-half-segments. The Ares 5 upper stage, known as the Earth Departure Stage, was designed to drive the Altair lander and the Orion spacecraft to lunar orbit; it would be powered by a J-2X rocket engine.

The NASA illustration shows how gigantic Ares 1 and Ares 5 are compared with the Space Shuttle. Ares 5, for example, has a huge shroud that could hold about eight school buses, and its rockets would pack enough power to boost almost 180,000 kg (396,000 lbs) into low Earth orbit. Ares 5 could in fact haul six times more mass and three times the volume of the Space Shuttle. Ares has the world's largest solid rocket motor. And even though the future of the Ares program is now in question, NASA will begin testing this motor in late 2010.

Ares 1 was designed to have a Launch Abort System (LAS), sitting on top of the Orion CEV capsule. The Launch Abort System is added to provide a safe escape for the astronaut crew in the event of an emergency during launch. The LAS was designed with four small solid rocket motors to allow the capsule to be jettisoned away from the launch vehicle. In 2010, after three years of development, the LAS was successfully tested. In the test, the unmanned Orion capsule shot more than a mile in the sky in just 20 seconds. Soon after, the parachutes opened and the capsule floated safely to Earth.

In mid 2010, the Orion crew capsule completed the Phase 1 safety review of NASA's human-rating requirements for space exploration in LEO and beyond. The capsule is on schedule to complete the critical design review in 2011, which could make the spacecraft ready for flight test as early as 2013, if funding is approved.

Ares 1 was also designed to be used for unmanned missions to the Moon to deliver payloads. Using a robotic lander, Ares 1 could carry payloads of up to 454 kg (998 lb) to facilitate transporting supplies and equipment needed for building the settlement, or to place a communications satellite into lunar orbit. NASA engineers also considered using Ares 1 for interplanetary missions.

Even if the plans of the Constellation Program are abandoned completely, the technology to safeguard the astronauts, the Orion capsule, the rocket motors, and many other aspects of the spacecraft design already developed can be adapted to future launch systems.

∝ ∞ ∝

When people go once again to the Moon, an interesting question to ask is, *what would they do there*? Many scientific activities are planned and many more ideas are as yet conceived. The NASA return to the Moon program of 2009 had two objectives: to carry out scientific research, and to prepare for human exploration of Mars. In the scientific pursuits, astronauts were to study the lunar terrain and test the crust to learn about the Moon's origin and evolution. They were to study the possibility of planet mining and of using the lunar natural resources for living. There are also plans to erect telescopes to support the search for alien life in extrasolar planets.

Where should astronauts build the settlement? This is another crucial question to answer since the selection of the site is the keystone to lunar planning and drives any lunar exploration program. The selection depends on several factors, including accessibility of the landing site, safety, exposure to sunlight, and availability of *in situ* resources. For example, building the base on a region that has permanent sunlight would ease daily operations and facilitate the generation of power using solar energy. Building on the top of a crater rim would allow more efficient power generation since the solar cells would be in constant sunlight.

There are several types of natural resources on the Moon: water, oxygen, and minerals. In addition to finding a multitude of interesting geological features to study on the lunar surface, explorers could extract nitrogen and calcium from the regolith—the lunar powdery rock surface. Researchers claim that the grainy substance could yield hydrogen-free silicon to make glass and ceramics that would be structurally superior to any made on Earth. The lunar soil also holds titanium, iron and aluminum.

Some years ago, the American space probe Clementine discovered a frozen lake in a large crater on the Moon's South Pole. Now, after analyzing Clementine's data, scientists concluded that there could be as much as 10 kg of ice per kg of rock. And more recently, other researchers revealed preliminary data from NASA's LCROSS lunar crater satellite, indicating that water exists in a permanently shadowed lunar crater. The discovery adds more possibilities to the plans for colonizing the Moon. The lunar base could be built nearby, as it would allow astronauts to liquefy the ice from the lake to make water for their personal use. Also, since water is made of hydrogen and oxygen, it can be processed to make the rocket propellant for future spaceships that would take off from the Moon on their way to other parts of the Solar System. Oxygen could be extracted from the regolith so abundant in the equatorial mare regions.

Initially, the settlement must be built with inflatable modules because they are lightweight and easier to transport from Earth. Several years ago, NASA began a study using the cold, harsh, isolated environment of Antarctica to test one of its concepts for astronaut habitats on the Moon, including a prototype inflatable tent to see how it fares during a year at McMurdo Station.

Other engineers suggested building the base structures buried within the regolith to protect astronauts from the deadly solar flares, and also to equalize internal and external pressure on the walls. Recently, NASA's Lunar Reconnaissance Orbiter (LRO) beamed back incredible images of caverns hundreds of feet deep. Scientists believe that the giant pits could be entrances to caverns, or skylights that formed when the ceilings of underground lava tubes collapsed. The caverns are huge, large enough to house a lunar settlement, and with their tunnels, they could offer a perfect radiation shield. Moreover, the caverns would provide a very benign thermal environment for the lunar explorers. Going about 2 meters under the surface of the Moon the temperature remains fairly constant, probably around -30 to -40 degrees Celsius. It is cold, but this temperature would be better than the extreme temperatures on the lunar surface that, as we said earlier, can range from a hot 100°C during the day to a frigid -150°C at night.

Any lunar exploration plan will undoubtedly include astronomy and astrophysics research. Telescopes built on the far side of the Moon could scan the heavens without the atmospheric veil that exists here on Earth. Some astronomers have suggested building an ultraviolet telescope that could use the rotation of the Moon to hunt for Earth-like exoplanets. These and many other ideas are proposed for the research that can be done on the Moon.

We know the Moon is a world, but not like ours. For instance, on the Moon there is no air for breathing. Although there is a thin atmosphere, it is relatively insignificant and tenuous in comparison with Earth's atmosphere—less than one hundred trillionth the density of ours—that we consider this layer of gases as non-existing. Thus, any human being on the Moon must wear a spacesuit to survive.

It will be very difficult for the human race to become entirely self-sufficient on the Moon, or in any other planet for that matter. But it is possible for humans to establish scientific research outposts, refueling stations, refineries, mining stations, or astronomical observatories. Ultimately, the best outcome of building a lunar settlement is that human beings would learn to live in another world and could test their skills to prepare them to continue exploring the Solar System. For now, however, we must consider the new plans of exploration that include human missions to the asteroids before we attempt the voyage to Mars.

To the Asteroids!

The asteroids emerge now as the next frontier in the human exploration of space. After cancelling the Constellation Program the president of the United States introduced a revised vision for NASA: to develop a new launch system and a spacecraft capable of going farther than LEO and land on an asteroid by 2025. This objective is much more ambitious than the original one of returning to the Moon. Traveling to an asteroid will be a very dangerous mission, perhaps the most risky and uncertain that could be attempted. The undertaking will be like an action-packed Hollywood movie.

In 2006, NASA's Constellation Program championed a study to examine the feasibility of sending astronauts to a near-Earth object (NEO). NEOs are asteroids or comets that are at perihelion distances less than or equal to 1.3 Astronomical Units, and can have orbits that cross that of the Earth. The mission was a sample collection voyage to a NEO in heliocentric orbits similar to Earth's, i.e. low inclination and low eccentricity. The astronauts would travel on the Orion Crew Exploration Vehicle (CEV) or similar spacecraft.

NASA's Near Earth Object Program has discovered over 5500 NEOs. Of these, nearly 1000 are classified as "potentially hazardous objects" (PHOs) or objects that pass within 0.05 AU of Earth's orbit. And these objects are very likely to be only a small fraction of the total number, as discovery rates increase each year. With so many cosmic rocks zipping by so near our planet, the probability is high that one of them will be—eventually— on a collision course with Earth.

NASA has not identified which among the hundreds NEOs would select for the first astronaut visit, but the main candidates are located at about 8 million kilometers (5 million miles) from the Earth. The Moon, in comparison, is much closer, at ~ 384,400 km, o less than a quarter of a million miles. Furthermore, the asteroids are small bodies compared with the Moon (our natural satellite has a diameter of more than 2000 km but an asteroid's diameter may be about 100 km or less), and thus they have very low gravity, likely much less than that of the Moon, making it very difficult and problematic to land on one. Astronauts would need an effective way of holding on to the rock with very resistant cables so that they don't float away in space. Once on the surface, astronauts would have a hard time walking on an asteroid.

A trip to a NEO may take about 200 days, while the round trip to the Moon requires just over a week. Nonetheless, the proponents of a mission to the asteroids emphatically state that a human expedition to a NEO would demonstrate the flexibility and utility of the Orion CEV and Ares or any other launch system that is selected in the near future. It would also be the first human expedition to an interplanetary body beyond the Earth–Moon system. Unquestionably, such deep space mission will be a most perilous undertaking, presenting unique operational challenges for the astronauts, for the spacecraft propulsion system, and also for the mission control team on the ground. However, NASA believes that performing a few piloted NEO missions will enable astronauts to gain operational experience in deep space, which will be necessary prerequisites for the eventual human missions to Mars.

In spite of the technical challenges of such mission, there are good reasons that justify the manned trips to the asteroids. Exploration of space, after all, has the main goal to learn about our Universe and knowledge is the main reason we attempt the most strenuous and grueling expeditions to foreign, inhospitable space environments. For example, the chemical composition of asteroids can give scientific evidence to understand the formation of planets and provide other clues to answer questions that are today unanswered. Scientists would gain new insights into the origin and history of asteroids, and also learn more about the evolution of our Solar System. The characteristics and contents of an asteroid can reveal something about the conditions and makeup of the solar nebula where the asteroid formed. That is why astronauts would need to collect material samples to bring back.

A mission of exploration to the asteroids would be a training field for humanity before attempting the trip to Mars, given the distance and the foreign atmosphere in that interplanetary space. If human beings cannot make the trip to a NEO then they may not make the trip to Mars.

A very important reason for undertaking such risky mission has to do with our survival on Earth. If astronauts can travel to and land on an asteroid, they could destroy one or change its trajectory when discovering that it moves threateningly towards the Earth. Although some researchers predict that the asteroid 2004 MN4 (renamed "99942 Apophis") has a 1 in 40,000 chance of impacting, none of the NEOs is right now on a collision course with our planet. However, asteroids have struck before and have left evidence of their impacts on the ground.

The 1.2 km wide Arizona Meteor Crater, for example, is believed to be the result of a cosmic object crashing on Arizona about 50,000 years ago. Another example, perhaps the most famous, is the Chicxulub Crater on the Yucatan Peninsula (Mexico) believed to be the result of the impact of an asteroid with the Earth about 65 million years ago. This impact is widely theorized to have been the event that exterminated many species. In more recent times, the atmospheric entry of a NEO was observed in the Tunguska region of Siberia, Russia. It is reported that on June 20, 1908, people saw the blazing object streaking the sky from 500 km away and the explosion was heard at a distance of 1270 km. The explosion is estimated to have been approximately the equivalent of 15 megatons of TNT.

So, if asteroids have struck before, they can strike again. Early in 2010, the news of a giant asteroid hitting the Earth created a sensation. The asteroid, which was discovered for the first time in 2004 and estimated to have the diameter of 1 km, was rediscovered by Russian astronomers. They claimed that the dangerous asteroid is revolving round the Sun, and although they could not estimate the real threat level, the scientists theorize that around 2028, when the asteroid moves closer to Earth, it could change its orbit and impact our planet. If it does happen, the future of humankind and other forms of life will be at dire risk. An asteroid impact could spell disaster and trigger massive natural calamities like tsunamis and earthquakes. It can also provoke significant changes in the climate.

Researchers in many parts of the world have studied methods for coping with such catastrophic event. For example, they have considered detonating nuclear or conventional explosives in, on or near the NEO, or using a tug of some type, whether connected to the NEO or using gravity to pull the NEO. Others are attempting to determine the effectiveness of the attachment of a long tether and ballast mass to an Earth-bound NEO to alter its trajectory. All these mitigation methods would require human intervention. If you saw the 1998 film "Deep Impact" you may have an idea how hard and complicated such maneuver would be. In the movie the heroes were after a comet, but the risk, technological challenge and arduous operation may have been realistically represented.

An interesting question to ask is which type of spacecraft would be more effective to transport the astronauts to an asteroid. One proposal is to use Orion or similar space capsule, and now the idea of using a segment of the International Space Station is gaining more momentum, as the end of its mission on LEO looms near. Yes, incredible as it may seem, the Space Station is slatted to retire in 2020. Many people have pondered what to do with it afterwards, as it is hard to imagine abandoning such as marvelous engineering structure in space just as it was done with Skylab in the 1970s. We cannot imagine that such as massive structure that took years to develop and construct would be left to drift in LEO until it eventually would decay and reentry.

And indeed, NASA is considering using part of the Space Station to build a spacecraft that would be sent to an asteroid, while also mulling more exotic artificial-gravity designs. A crew compartment, like the Tranquility node for example, could be modified for such a journey. The proposals at this stage are just concepts rather than part of a final design for an asteroid mission. However, plans for reusing any components of the Space Station.

What would astronauts do on a first sample return mission to a NEO? They would test several different collection techniques and target specific areas of interest via extra-vehicular activities (EVAs) more efficiently than robotic spacecraft. Studying NEOs is vital to understanding the evolution and thermal histories of these space bodies during the formation of the early Solar System. The data that astronauts collect would help determine what materials may have been delivered to the early Earth, and would help determine the origin of NEOs. In addition, astronauts could test designs for extraterrestrial resource extraction systems and try their utilization.

Researchers claim the mission to a NEO would be less expensive than a trip either to the Moon or to Mars. Undoubtedly, a mission to NEOs would provide a great deal of technical and engineering data on spacecraft operations for future human space exploration to Mars and beyond while conducting scientific investigations of these space objects. Nevertheless, there are still a number of issues that need to be resolved such as the potential high radiation the crew will be exposed to during deep space travel, requiring that effective radiation mitigation strategies be in place before such interplanetary mission is carried out.

New navigation technologies are also needed to rendezvous with, and land upon, a relatively small asteroid. A spacecraft will need intelligent navigation, guidance and control systems to safeguard the crew even further. And although a highly successful automated mission by the Japanese Hayabusa spacecraft is providing a wealth of data to help improving on future sample return missions to a NEO, additional characterization of the target asteroid is required for human mission planning and crew safety so an additional robotic mission is recommended.

If human beings learn to manipulate the asteroids (destroy or deflect their collision path), then they could save the Earth from the danger those cosmic projectiles pose when they move close on a collision trajectory with our planet. A window of opportunity for a crewed NEO Mission may exist in the ~2015-2030 timeframe. Will we be ready?

To Mars and Beyond

The human exploration of Mars will be much more complex, both in scale and in distance, than any other space mission attempted to date. Yet, the prospects for establishing a human presence in that planet are exhilarating, and the fate of human evolution makes it necessary that we begin planning now. A month after the Phoenix Martian lander arrived, researchers at NASA's Jet Propulsion Lab (JPL) announced that the Martian soil could potentially support plant life, suggesting it may be possible to grow vegetables and fruits to feed future explorers.

Phoenix also discovered clays and calcium carbonate in the Martian soil; these compounds could only be made with liquid water, suggesting (according to some scientists) that the Polar region of Mars might have supported life in the past when the planet was warmer. Laser instruments aboard the Phoenix lander also detected snow in clouds floating 2.5 miles above the surface. And in the last few years, biological or geological activity was detected. A substantial plume of methane seemed to float through its atmosphere implying that the planet is teeming with life.

To build a settlement on the Red Planet will require multiple trips with robotic spacecraft to send supplies and the equipment needed to set up the facilities before the explorers arrive. Initially, it is expected that six to eight individual launches would be required to assemble the material that could be sent to Mars.

NASA is considering chemical rockets for the trip to Mars for one reason: this propulsion technology is sufficiently mature, ensuring the availability of reliable rocket engines with verified performance. The spacecraft would need two powerful bursts of energy—one to attain mid-Earth orbit, and one to continue into interplanetary space. For now, this is only possible to do it effectively with chemical rockets.

The long voyage to Mars will need more propellant—to go and return—thus requiring larger propellant tanks. The crew will need more room to live during the longer trip. This in turn may result in a much larger interplanetary spacecraft. That is why the initial stages (including engine and propellant tanks) of the Ares launch vehicle would be discarded after the propellant is exhausted during ascent. Once in Earth orbit, the remaining upper stage that contains the crew capsule, the Mars orbiter, and the exploration vehicles would keep going on its voyage to the Red Planet.

There are many diverse ideas being pursued at this time for the human journey to Mars. The British company Reaction Engines Ltd., for example, has devised a very believable Mars mission scenario where its spaceplane SKYLON would be used as the launch vehicle to Earth orbit. They developed the Troy mission concept as a result of a feasibility study performed to confirm the capability of the SKYLON launch vehicles to enable large human exploration mission to the planets. To showcase the Troy program, the company developed a movie that shows how a feasible, affordable, practical and safe mission to Mars can be achieved using SKYLON, enabling a team of explorers to land at three sites and explore over 90% of the surface. This mission architecture also creates the infrastructure to support further human missions. The Project TROY movie shows what a Mars mission would look like and how a settlement would be constructed.

According to the company's description on its website, the Troy mission could be performed in two parts—an unmanned, precursor mission, and the later manned or piloted mission. Using SKYLON, the components for the Troy ships would be delivered to an Orbital Base Station, where they would be assembled. These components would all fit into the SKYLON payload bay, and have as much commonality as possible. The Precursor Ships include an Earth Departure Stage, a Mars Transfer Stage, three large Landing Modules, (one carrying rovers, one carrying power and propellant production plant, and one carrying an accommodation module), and a Ferry Vehicle for the journey to and from Mars orbit to the surface. Transit time to Mars would be 264 days, and on arrival, the three landing modules would be deployed to land at a pre-selected site, forming a base. Three Precursor Ships would be sent, forming three bases on Mars which would enable the rovers to reach more than 90% of the Martian surface.

The Manned Ships would include Mars Transfer Stages, an Earth Return Stage, crew accommodation modules, and capsules for the return to Earth orbit. These piloted spacecraft would be sent off at the next launch opportunity, and take 251 days (more than 8 months!) to reach the Red Planet. Upon arrival, the crew would dock with the waiting Precursor Ships and the astronauts would transfer to the Ferry vehicles for the trip to the Martian surface, where they would land at the pre-prepared bases. The astronaut teams, each comprising 6 people, would spend 14 months exploring the Martian surface, making extensive use of the rovers.

Meanwhile, atmospheric processors at each of the three bases would collect gases, separating and storing them for use by the explorers and to provide propellant for the ferry vehicles. At the end of the exploration phase, the crew would use the Ferry vehicles to leave the surface and dock with the orbiting missions ships, and from there begin the journey back home. Once in the vicinity of the Earth, the return capsules would separate and perform an aerobraking maneuver to slow down and dock with waiting SKYLONs for return to Earth.

Very nice scenario, don't you think? It seems simple and attainable, and all that is left to do is develop the required technology and the expertise of humans to put it in practice. However, it may take decades or a century before humans gain the necessary experience to live off the land for 14 months in an extraterrestrial world. Thus far, humans have lived in low Earth orbit for months at a time, but they have access to food, water and other life-support technology in the Space Station. In case of emergency, they can return home within days. In the case of a mission to Mars, it'd take months to make the trip back.

Human exploration, for the foreseeable future, must be concentrated on the Moon and the asteroids. After gaining experience both to travel farther then LEO and to live in extraterrestrial environments, then a mission to Mars makes sense. Beyond that, the next step in our human exploration of space could be trips to the large moons of the outer planets. For those long interplanetary trips beyond Mars, chemical propulsion is no longer an option. We will need to advance de development of nuclear rockets, plasma rockets, antimatter rockets, and maybe other propulsion types as yet considered.

A few years ago NASA assessed the possibility of using liquid methane rocket engines to power the spacecraft to Mars. These are rockets powered by methane (CH_4) fuel mixed with liquid oxygen, a very attractive option for future interplanetary missions. The methane rocket operates in the same manner as any conventional hydrogen/oxygen rocket does. But liquid hydrogen requires large and heavy, highly insulated tanks because the liquid hydrogen must be stored at very low temperatures, about -252.9°C. Methane, on the other hand, can be stored at warmer temperatures, so the tanks would not require that much insulation and would be lighter. The tanks would also be smaller, as liquid methane is denser than liquid hydrogen.

If there was fuel at the destination, a space vehicle leaving from the Earth would not have to carry so much propellant. With fuel depots at its destination, the spacecraft would travel lighter and farther. That is why methane is also attractive as a rocket fuel; it is abundant in the outer regions of the Solar System. Methane could be harvested from Mars, Jupiter, and other planets and their moons. A sample-return mission to the outer planets in the Solar System—something that has never been attempted—could be achieved with methane rockets, if one considers the possibility of building fuel depots at Titan to refuel spacecraft traveling farther away.

Titan is the largest moon of Saturn, the only natural satellite known to have a dense atmosphere. It was discovered in 1655 by the Dutch astronomer Christiaan Huygens. This moon has a diameter roughly 50% larger than our Moon and is 80% more massive. Most importantly, we have evidence that Titan is the only object other than Earth to contain liquid hydrocarbons that rain from the atmosphere into lakes and rivers.

It may sound like science fiction but the idea of a space vehicle powered by methane rockets landing on Titan, gathering samples and refilling its fuel tanks to return to Earth is not as farfetched as it may seem today. But we need many more years to develop all required technology.

In the case of Mars, methane fuel could be manufactured there. To realize this idea it will require first building a Mars settlement and then constructing a fuel factory. The factory would collect carbon dioxide from the Martian atmosphere and then mix it with hydrogen that could be processed from Martian ice. Heating the carbon dioxide-hydrogen mixture produces methane and water. And there you have it: fuel for the rockets and water for the astronauts on the Mars settlement.

In addition, other forms of propulsion that don't use chemical energy have also been considered. Extensive studies of nuclear thermal rockets and fusion rockets promise to reduce the trip time to Mars to less than six months. Plasma propulsion engines are also very promising. In fact, a prototype plasma rocket, better known as the Variable Specific Impulse Magnetoplasma Rocket (VASIMR), has been developed to the point that is ready to be tested in orbit in 2012.

The VASIMR uses RF energy to heat a gas to extreme temperatures to make plasma that is ejected from the engine at very high velocity, resulting on a very high specific impulse. It uses powerful magnets to shield the engine structure from the superhot plasma. NASA and the developers of the advanced plasma rocket concept plan to place a 200-kilowatt version of the VASIMR on the Space Station ISS to test its performance in space. The VASIMR will draw power from the station solar arrays to charge batteries that in turn will drive the high specific impulse plasma rocket.

Many other ideas for propelling a spacecraft are currently studied. NASA's goal has always been to create breakthrough technologies to help transport astronauts to any extraterrestrial destination. For example, in the next few years NASA will conduct with DARPA a study that will explore power beam propulsion, a technology that could direct a beam of energy at a spacecraft to power its flight, either to launch from Earth or to accelerate through space.

Nuclear electric propulsion is another concept considered and studied since the 1960s. One such study came from NASA researchers who considered the prospect of sending humans to Callisto, one of Jupiter's moons, using an all Nuclear Electric Propulsion (NEP) space transportation system. The recommended NEP architecture would use magnetoplasmadynamic (MPD) thrusters. The mission timeframe assumed on-going human Moon and Mars missions and existing space infrastructure to support launch of cargo and crewed spacecraft to Jupiter in 2041 and 2045, respectively.

Why Callisto? It is the third largest satellite in the Solar System, and the outermost Galilean moon of Jupiter. Orbiting at a distance of ~1.9 million kilometers, Callisto is located beyond Jupiter's main radiation belts making its local environment more conducive to human exploration. Callisto is an icy, rocky world with a surface gravity of ~0.127 g_E and a composition consisting of water-ice and rock in a mixture ratio of 55:45. Besides having significant quantities of water-ice for propellant production, Callisto's heavily cratered and ancient landscape (~4 billion years old) suggests that significant quantities of non-ice materials and asteroid dust may be on its surface. Thus, scientists considered a human mission to Callisto as a worthy exploration destination beyond Mars mainly due to the availability of in-situ resources to support the mission.

For extended robotic missions such as sample return and asteroid rendezvous, electric propulsion systems have proven to be very efficient propulsion methods. Ion rockets, for example, have very high fuel efficiency and afford the spacecraft an unprecedented maneuverability not possible with chemical rockets: they can exit the orbit of one distant body, and fly to and orbit another.

Ion rockets ionize the atoms of the propellant gas (typically Xenon) and then, with a strong electric field, expel these ions at high speed. Ionization is done by microwaves (using solar energy collected through solar arrays) and acceleration with a high voltage to generate the required thrust.

One disadvantage of ion rockets is that the levels of thrust produced tend to be several orders of magnitude lower than those given by chemical motors. Thus, to achieve the same overall change in momentum the ion rocket must operate for longer and must therefore be more reliably.

Mission analysis is more complex with ion rockets than is the case for a traditional chemical propulsion system. Instead of the spacecraft firing its engines for relatively short periods of time like chemical rockets (but using vast amounts of fuel in doing so and running out of it quickly), to achieve the same momentum change, an electric propulsion system has to operate for far longer, leading to a burn that can last for a significant fraction of the spacecraft's orbit. But this is very advantageous for interplanetary unmanned missions. Dawn, for example, the robot on its way to explore the asteroids Vesta and Ceres, is powered by a very efficient ion rocket.

Typically, spacecraft powered by conventional rockets coast to their destination to conserve fuel. Dawn's ion engines, on the other hand, are almost constantly active. It will thrust for 5 ½ years! According to NASA's chief engineer, Dawn has already been thrusting for 591 days. That's 62% of the time it has been in space. Dawn uses only a kilogram of xenon every 4 days.

For human missions, we need advanced propulsion concepts to reduce the spacecraft's mass and increase its spaceflight performance for interplanetary trips. However, the technology to support the design and construction of new propulsion systems for the large spacecraft necessary for a crewed mission, require much more development before it can be implemented. As of today, most missions to Mars suggest that it will take a minimum of 6 month to make the trip with available technology and with optimized trajectories for the interplanetary voyage. That's a long trip, indeed!

Space Navigation

There are considerable fundamental differences between spaceflight and any type of travel on out planet. The most important difference is in the use of energy. An airplane, a sea vessel, or any other kind of ground transportation will move as long as energy is supplied to their propulsion systems, and as soon as the energy supply is shut off, the vehicle will decelerate (by friction) and finally stop. In space there is no atmospheric friction, and once a spacecraft is set in motion, it will continue in this state of motion. A rocket-powered spacecraft must accelerate its propellant onboard, and this costs energy, so one would want to burn up all propellant as quickly as possible, and thus the greater part of the space voyage will take place in free-flight (coasting) without thrusting. This is fine and good for a robotic spacecraft on a one-way trip. But for human missions, we need to carry the propellant onboard to ensure we make the trip back home. The required amount of propellant for the return trip could be considerable.

Another difference of spaceflight is that the energy requirements are governed by the direction in which a spacecraft moves. It makes a huge difference whether the vehicle moves in the direction in which the Earth is moving around the Sun, that is, in the counterclockwise direction or in the opposite direction. For example, in order to escape from our Solar System, starting from Earth, a spacecraft requires in the latter case a characteristic velocity which is more than four times as high as for the first case. This is due to Earth's velocity around the Sun, which is on average 29.8 km/s. In the first case, this speed works on favor to the escaping spacecraft, but in the second case it works against it. Since all planets and almost all satellites and comets move in this direction, a spacecraft would move in the same direction around the Sun.

On the Earth, changes in a trip's direction can be carried out without much loss of energy. But in space, any change of direction will require a great deal of energy. That is why is so important that we determine as accurately as possible the direction along which a spacecraft will take in a given mission so that during the trip only small corrections in the flight path are needed. Also, the amount of energy required for spaceflight depends on the shape of the orbit. And as we said earlier, another feature of spaceflight that makes it so different to any other type of travel on our planet is that the starting point and the destination in space are not fixed points but rather they are bodies moving a different speed around the Sun. This makes space navigation considerably more difficult than any other navigation on Earth, where the points of departure and arrival remain fixed.

Another crucial difference of spaceflight is that, since there is no friction in space, we must always adapt the spacecraft's velocity to the velocity of the planet or moon of destination. It does not matter which of those two speeds is the greater, in both cases we have to eliminate the speed difference by using more energy. All this suggests that space travel will require much more energy: energy to set the spacecraft in motion and energy to stop it.

Furthermore, all motion in interplanetary space is governed by the Sun. Our mother star, whose mass is 332,000 times as large as that of our planet, and 1050 times as large as that of the largest planet, Jupiter, rules all motions within our Solar System.

One of the most important steps in designing any space mission is determining the trajectory that a spacecraft will follow in route to its final destination. In fact, the success and even the feasibility of any mission depend primarily on the design of its trajectory. This is even more important for a human mission.

In the study, two essential design parameters for the trajectory of a mission are considered: the overall time of the mission, including transit and eventual stays or orbits and the propulsion requirements to achieve the desired trajectories and orbits. These two parameters are critical in the design of any mission, for they affect the human factors (food, water, medical, supplies, psychology) and the hardware requirements (propulsion, engines, systems, etc.) of the mission.

Many trajectory options for the human exploration of Mars have been proposed through the years. Among the suggested options the most attractive for crewed expeditions are conjunction-class missions. These are missions characterized by short in-space durations with long surface stays, as opposed to the long in-space durations and short surface stays characteristic of opposition-class missions. Of course one has to also consider the return trip, and thus a trade off the Earth-Mars and Mars-Earth trajectories must be conducted to determine mission opportunities and transfer times in order to optimize a roundtrip.

By following the most optimized orbital path and starting from the Space Station, it will require a large velocity to make the round trip to Mars. Consider this: we need to escape from the orbit of the Space Station, with sufficient speed to start with finite acceleration. Then we need additional speed for orbit transfer between Earth's orbit and Mars' orbit—a large fraction of the total velocity—plus velocity to adapt to a circular orbit around Mars. And finally we need velocity to soft land on Mars. We can assume that on arrival near Mars we move around the planet in a circular orbit and that no energy is required to coast down to land. Under the most idealized assumptions, a total velocity for the outward voyage could be on the order of 10-11 km/s. The velocity for the return trip could be higher. We have carried out numerous calculations and estimate that the total velocity for the round trip to Mars is more than 23 km/s, which is far too large for chemical rockets.

Therefore we must continue developing rockets that use other types of energy. For example, nuclear energy is the next best choice. Nuclear forces are stronger than atomic forces. With nuclear energy must be possible to achieve a greater concentration of power than with chemical energy.

The trajectory study must also include the potential for aborting a mission without capture into Mars orbit either for the crew to return to Earth after a certain period of time, and for the case when the propulsive capability of the transfer vehicle is used to modify the trajectory during a Mars swing-by. As mentioned earlier, the propulsive requirements of a mission are related to spacecraft mass, and thus are very important to optimize trajectories for human missions to far away places such as Mars.

For a mission to Mars, a launch window occurs approximately every 26 months, when Mars and Earth reach a position in their respective orbits that offers the best alignment of the two planets. The next launch window for a mission to Mars is February 2012. But we won't be ready to go just yet. It may be a couple of decades before we do.

A trip to Jupiter, for example, is then left for the second half of the 21st century. The trip itself would take many months, even with a fast spacecraft propelled by a nuclear rocket. Think about it: when

Jupiter and Earth are closest, Jupiter is 628,743,036 km (390,682,810 miles) away. We would also need to develop other life-support technologies. Interplanetary trips may seem so easy in the movies, but the truth is, human beings cannot tolerate long periods of weightlessness, so gravity must be simulated onboard.

Final Remarks

Propulsion capability will continue to rule our ability to travel across the planets and beyond. Let us assume that in the next few decades we can develop a propulsion system capable of achieving velocities of 50-100 km/s. For interplanetary missions, this capability could be sufficient. But if we try to leave the Solar System, to reach the nearest stars, these speeds will not do. A simple analysis show us that a space vehicle traveling at 100 km/s would require more than 12,000 years to make a voyage to our closest star, Proxima Centauri, which is 4.3 light-years away.

Even moving at the speed of light, we cannot go far within the span of our lifetime. In our cosmic neighborhood, within a radius of 12 light-years, there are only about 18 stars, and that is only a tiny fraction of our Galaxy. Thus, we could only expect to explore the immediate neighborhood of our Solar System at light speed, unless we manage to manipulate space-time and find a way to circumvent the light-speed limit.

Spacecraft with artificial gravity systems must be in place together with protective radiation shields, to help thwart the negative effects of interplanetary travel on the human body. A spinning centrifuge can be used to generate artificial gravity. Furthermore, if the astronauts were put in a deep and monitored slumber, in a frozen state, they could make the long and lonely interplanetary journeys more comfortable. The idea of putting the crew into hibernation or animated state for deep space travel is no longer a topic of science fiction novels. In fact, the freezing of the human body is part of the medical research on Earth, for the purpose of saving lives. A study published in 2010 reveals that a frozen human body can be revived. The study revealed a capacity—that was unknown before—of the organisms to survive the lethal cold momentarily, slowing down the biological processes that sustain the life. This type of investigation is a necessary component of space medicine that support plans for space travel and colonization.

Just as we need to advance propulsion technologies to reduce the trip time, we also need to develop hibernation technology for humans, systems of artificial gravity with engineering methods to rotate large spacecraft in deep space, and effective radiation shields. These and other advanced systems to support human spaceflight require much more research, development and engineering, and I hope more scientists and engineers in many nations join their efforts so that we can make the voyages to the planets beyond Mars, in this century!

12
Living in Space

Was Einstein wrong about space travel? His theory of relativity tells us that the faster one travels through space, the slower one travels through time, making us younger, right? Did he take into account biology? How about the effects of radiation on the human body? I wonder if space travel would make us prematurely old…

Artist's concept of a lunar facility in the eastern *Mare Serenitatis.* Credit: NASA/Pat Rawlings (SAIC)

Like most living creatures on Earth, humans evolved at the surface, at the "bottom of the gravity well." Thus, our bodies are not suited for living in space. In space, there is no air to breathe, and an astronaut without her spacesuit would swell grotesquely and die.

On the ground, our bodies are subjected to the pull of Earth's gravity, but without gravity to work against, muscles waste away and bones become brittle as they lose calcium. It is reported that 3.2% of the bone loss occurs after just 10 days in microgravity conditions. Astronauts lose about as much bone mass per month in space as postmenopausal women with osteoporosis do in a year. The loss of calcium through urine may result in painful kidney stones, and the reduction in bone density leads to higher risk for bone fractures.

With the exception of trips to the Moon, most human spaceflights in the last forty years have been to LEO. Currently, the only people living in space up to six months at a time are the astronauts working on the International Space Station (ISS), which orbits at approximately 350 km above the surface of the Earth. At that altitude, the acceleration of gravity is about 90 percent of the gravity on the ground, and such reduction is enough to adversely affect the function of the human body.

The physiological and psychological effects of space are many and varied. Adverse physiological effects include effects on the cardiovascular system, on the bones, on the muscles, space sickness, effects on the immune system, and the radiation of space has a high potential for causing cancer.

In microgravity conditions, astronauts develop "space legs," that is, the legs become thinner, and the calves become smaller because the leg muscles force blood and other body fluids upward. With the fluid accumulating in the head, the face becomes swollen and flush. Upon returning to Earth, the fluid distribution in the body changes quickly, provoking a tendency to develop hypotension—abnormally low blood pressure.

Without the force of gravity the human body floats. And thus astronauts move by pressing gently against the walls of the spacecraft. And in spite of the many hours of exercise they perform, long-term exposure to microgravity results in muscle deterioration, and other physiological changes requiring significant rehabilitation once the astronauts return to Earth.

Former astronaut William Pogue wrote about his experience, giving a vivid image of the effects he suffered living in space. In his 1992 book ("*How do you go to the bathroom in space?*"), Pogue wrote:

> *The first thing you notice when you go into space is an absence of pressure in your body. You may feel light headed or giddy. After a half hour or so, your face may fee flushed and you might feel a throbbing in your neck. As you move about, you will notice a strong sensation of spinning or tumbling every time you turn or nod your head. This makes some people uncomfortable or nauseated. You will also have a very "full feeling" or stuffiness in your head.*

You may get a bad headache after a few hours, and this too may make you feel sick to your stomach.

In addition to getting a reddish, swollen face and thinner lower legs, astronauts have a problem with spatial orientation. The reduced gravity affects humans' perception of size and distance. Objects may appear longer and farther. Although this condition appears to be psychological, it is still a problem that requires correction, especially since astronauts are required to perform many tasks that requires high precision. In orbit, the concepts of "up" and "down" are abstracts, and so living in a space station can be bewildering. Astronauts report that whenever they go through a doorway from one module of a spacecraft into another, they have to pause to decide which way to turn their floating bodies. It seems fun when we see astronauts doing zero-G stunts on the International Space Station. But you should know that they can only do those summersaults only after the urge to vomit from space sickness has faded.

Dizziness is common in space and it has to do with the mixed messages that the brain receives from the different parts of the balance or equilibrium system. Sometimes the eyes, ears, the skin and the muscles have different sensations. The eyes may center in one point, while the body is actually floating around. As a result, the brain is not really sure where the body is in space. This causes nausea and dizziness. Fortunately, these feelings disappear after a few days.

Researchers think that studying animals could help understand how to overcome space sickness. For example fish, snails, and toadfish could be used to determine whether living in space can create long-term or even permanent damage in the inner ear. The inner ears of these animals are very similar to humans', and so they could be studied in space to determine how they can adapt to zero gravity. Fish might be good candidates for longer studies and thus could be sent in a spacecraft to Mars or beyond with a return ticket, given that they can live for 40 or 50 years.

In space the concept of day and night is a bit abstract. For example, every 24 hours, astronauts on board the ISS will experience 15 dawns as the station speeds around the world. That is why the work schedule of the crew in the ISS allows them to work and sleep to fixed hours that match their circadian rhythms. And although the astronauts work many hours they also dedicate time for physical exercise to maintain their muscles strong.

LIVING IN SPACE

Space travelers need special spacesuits. In outer space there is no atmosphere and thus no atmospheric pressure and no oxygen to breathe; therefore, human beings must take their environment with them. On the ground, Earth's atmosphere is about 20 percent oxygen and 80 percent nitrogen. If you were to climb a mountain, at an altitude of 5500 m you'd find that the atmosphere is half as dense, and at higher altitudes greater than 12,000 m, air is so thin and the amount of oxygen so small that you could not breathe.

So, at an altitude above 19,000 m, humans must wear a spacesuit that can supply oxygen for breathing and that can maintain a pressure around the body to keep body fluids from boiling. For safety reasons, astronauts wear a pressurized spacesuit during launch and landing. They remove the bulky spacesuit when they arrive to the pressurized ISS. But astronauts must wear the spacesuits when they work outside the space station—performing space walks or extra vehicular activity (EVA). They must also wear spacesuits to live on the Moon or on another planet where there is no atmosphere either.

During EVA astronauts are exposed to the rigorous ambient of deep space, and even wearing their space suits cosmic radiation can be very dangerous. On the Earth, the atmosphere and the magnetic field that surround it act as a radiation shield, stopping dangerous radiation and thus protecting us. However, this protective shield is absent in space and the astronauts are exposed to it.

Astronauts on EVA face great danger. If their pressurized spacesuits would rip while working on the repair of a system outside the spacecraft, it would cause the death of the astronaut within minutes. EVA is a very difficult activity. Astronauts require many hours of training to be able to perform EVAs. They attend classes on technical and theoretical aspects of the mission and perform simulations of procedures to be used during the mission. The simulations are conducted in an environment as similar to space as possible—a swimming pool. This is a very special swimming pool, where astronauts learn to perform complex maneuvers and repair or build parts of the spacecraft while wearing the cumbersome spacesuits.

Comfort—or lack of comfort—is another issue. Whether orbiting onboard a space station or voyaging to some destination in space, astronauts must adapt to live in very uncomfortable conditions. As one blogger wrote, "Tomorrow's Astronauts Will Fly Economy Class to the Moon." That was a clever commentary after NASA revealed Orion, announcing that it was its next-generation spacecraft to the International Space Station and (eventually) to the Moon. Orion is the 5-m diameter Orion Crew Exploration Vehicle, very similar to the Apollo capsule. There is no much room to move around, and so astronauts must be strapped to their seats for long hours.

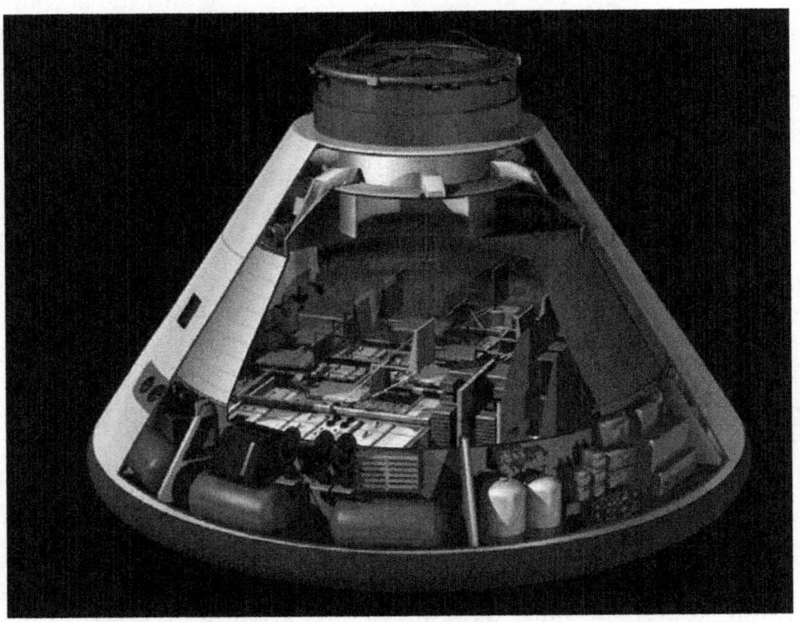

Orion basic configuration, showing 6 seats – Credit: NASA

On the more spacious International Space Station (ISS), astronauts sleep on sleeping bags strapped to the wall of the living module, eat dried-frozen meals on the go, and sip their beverages from plastic squeeze-drink bags, every day!

Living conditions on any spacecraft can be harsh. Yet, In spite of all the difficulties and adverse effects of the inhospitable space, human beings have adapted remarkably well to the conditions of reduced gravity. Since 1961, more than 450 men and women they have gone and lived in space. Astronauts, cosmonauts, taikonauts and space tourists of more than 33 nations have traveled in Soviet, American,

Russian and Chinese spacecraft, demonstrating that the human body is able to overcome many challenges and to adapt to other ways of life.

Living on the Moon

The first human missions to the Moon were brief. In the future, space explorers will go to stay and will experience harsher conditions compared with those felt by the Apollo astronauts. Future lunar explorers will spend longer periods of time building the bases and eventually living on the Moon for extended periods of time. Eventually, they will learn how to inhabit other worlds in the sky. But it won't be easy, and it may take generations to adapt the human body to those extraterrestrial conditions.

The Moon is a world unlike ours, desolate and barren. The footprints left behind by the Apollo astronauts may still be there because there is no wind to disturb them. There is no atmosphere, and the temperatures are extreme: it is very hot during the day (100 °C) and very cold at night (-173 °C). The surface gravity on the lunar surface is 1.63 m/sec^2, a fraction compared with the surface gravity of the Earth (9.8 m/sec^2).

The Moon's "atmosphere" is made up of gases such as radon and helium that originate from radioactive decay within the crust and mantle. Another component of the gases over the lunar surface develops from bombardment by micrometeorites and the solar wind, the energized flow composed of charged particles that originate in the Sun's atmosphere.

Astronauts on the Moon will be exposed to direct radiation from solar storms, as there is no protective atmosphere to shield them from the harmful rays. On the Earth we are protected by the magnetosphere which deviates many of the harmful cosmic particles; the terrestrial atmosphere absorbs other energetic particles. But in space or on the surface of the Moon where there is no protective atmosphere, a major eruption of the Sun could be deadly. A solar burst, which is a violent release of energy, could enclose an astronaut and everything around her burning her to a crisp. Apollo astronauts were lucky, since they went to the Moon for just a few days during a period of solar tranquility. The explorers of the future will live in the Moon for a long time.

In addition to the adverse biological effects we mentioned earlier, another hazard of living in space is meteoroid bombardment. Explosions caused by meteoroids crashing into the Moon, for example, are common and can be dangerous to spacecraft and astronauts. Meteoroids that hit Earth disintegrate in the atmosphere, producing a harmless streak of light. But the Moon has no atmosphere, so lunar meteors plunge into the ground and cause damage. Impact velocity of meteoroids is between 12 and 72 km/s; compare it with the impact velocity of a gun bullet which is just 0.34 km/s.

Typical strikes on the Moon release as much energy as 100 kg of TNT (Trinitrotoluene, an explosive), gouging craters several meters wide and producing bursts of light bright enough to be seen on Earth through ordinary telescopes. The potential for meteoroid hitting spacecraft, colony buildings, and future Moon explores is now being evaluated. It is estimated that a 1-mm diameter meteoroid could make a hole in an astronaut's spacesuit.

The lunar dust will be more than a nuisance for the explorers. As the Apollo astronauts discovered, the surface of the Moon is covered with a powder layer between four inches to a yard deep. This pulverized blanket was produced by a few billion years' worth of micrometeorites slamming into the lunar surface. Anything with moving parts would be affected by the electrically-charged dust. The light-reflecting scientific instruments placed on the Moon by astronauts 40 years ago have mysteriously degraded. The degradation is likely due to lunar dust. Equipment placed on the barren, weatherless lunar surface can in fact suffer performance problems in the long term.

Adverse Effects of Interplanetary Space

As we expand our presence in space and travel farther, the physiological and psychological effects on the human body will be more severe. The long voyage to Mars will be very challenging. During the cruise through deep space, lasting several months, the crew will be exposed to the effects of radiation: protons from solar flares, gamma-rays, and cosmic rays from exploding stars. For example, during the trip across the Solar System, the Voyager spacecraft encountered radiation levels at Jupiter that would be 1000 times the lethal level for a human being.

The radiation risks on the human body associated with interplanetary voyages have yet to be fully quantified. Scientists estimate radiation danger in units of cancer risk, based on assessments of Hiroshima atomic bomb survivors and cancer patients who have undergone radiation therapy. On Earth, a healthy 40-year-old non-smoking American male stands a 20% chance of eventually dying from cancer. What is the risk when exposed to the environment of deep space?

According to a 2001 study of people exposed to large doses of radiation, the added risk of a 1000-day Mars mission lies somewhere between 1 and 19 percent. Some researchers think that the answer may be about 3.4 percent. The odds are even worse for women. Because of breasts and ovaries, the risk to female astronauts is nearly double the risk to males. But if the added risk of radiation exposure is 19 percent, scientists predict that an astronaut would face a whopping 39 percent chance of developing life-ending cancer after returning to Earth. Not a good prospect for space explorers, unless they are protected with radiation shields.

In the NASA study, researchers assumed the interplanetary vehicle would be built mostly of aluminum, like the old Apollo command module. The aluminum skin of the vehicle can absorb about half the radiation. Aluminum is an excellent material for spaceship construction, because it is lightweight, strong, and familiar to engineers from long decades of use in the aerospace industry. Radiation shields could reduce the exposure for astronauts even more, and engineers continue developing better concepts.

The International Space Station was designed to have a thick-walled room where the astronauts can hide during times of increased solar radiation. Better concepts are needed for interplanetary spacecraft. British scientists have proposed to use magnetic shields that could be deployed around spacecraft and on the surfaces of planets to deflect harmful energetic particles. This concept would mimic the magnetic field which protects the Earth. It consists of system that would be about the size of a "playground roundabout" and use a tiny amount of energy, using two mini-magnetospheres housed in two outrider satellites in front of the spacecraft. The researchers tested the system and found that it offers almost total protection. This method of radiation shielding may be good to protect the crew in future interplanetary missions.

The psychological effects of spaceflight are equally challenging. If you have taken a trip on a commercial jet plane that took eight or ten hours, remember how confined you felt, strapped to your seat and with only a few bathroom breaks to stretch your legs. In those ten hours you probably read, watched videos, played games, ate, slept, and you may have looked through the small windows of the aircraft to enjoy the views of the sky, the clouds, or the terrain on the ground below. The thrill of going on vacation was perhaps replaced by boredom or anxiety, not being able to get off the plane until it landed.

Well, imagine you are traveling to Mars sharing a spaceship with other astronauts. As weeks go by, you and everyone in the capsule will experience extreme isolation and confinement. What will you see through the portholes of the spacecraft? The inky darkness of interplanetary space may be overwhelming. How will you feel when you lose sight of our planet? There are no stopover stations between Earth and Mars. No instant message is possible being far in deep space. If you want to call someone back home, you'd have to wait a long time to get a reply. A radio signal could take 40 long minutes or more to travel to controllers on the ground and then back to you.

The interplanetary spaceship will be a tiny little world where the crew will live for several months, an extremely long time to be so confined. The food will be rationed. The water will be continuously recycled. Every drop of bodily fluid expelled from each member of the crew will be processed and re-ingested. Habitable areas will be very small. I am sure astronauts will miss the comforts of home and the views of our planet. Such isolation will have psychological consequences, affecting the mental state of the members of the team living so close together, and knowing they cannot escape and return home right away.

Experience obtained in previous space missions has revealed that conflicts develop among the members of the crew 30 days after launch, showing hostility against others and having frequent disputes. Conflicts between the crew and the mission control team on the ground are also common. Thus, during an interplanetary voyage, the psychological support for astronauts will be imperative. The crew will need long distance counseling and regularly scheduled communication with members of their family using bidirectional voice and video communication. But in long trips it will not be possible to have instantaneous communication with counselors when a problem arises.

Team dynamics will be crucial to the success of the mission. Each one of the crew members would have to rely on all others for survival. They will have to make real-time decisions about daily operations, without input from mission control on Earth. The success of such interplanetary mission may depend to a great extent on how you get along with each other, and on the mental endurance of each astronaut to overcome many unforeseen challenges.

Studies on the ground attempt to simulate the team dynamics of a long space voyage. In the spring of 2009, the European Space Agency (ESA) and Russia's Institute of Medical and Biological Problems teamed up to carry out an experiment to simulate a manned flight to Mars of 520 days. Ten candidates were initially selected, and in March 2010 the final crew was reduced to six. The volunteers of the experiment identified as Mars 500 include three Russians, two Europeans, and one Chinese. All male.

Mars 500, started on June 2010, will test many of the scenarios anticipated for the long interplanetary trip, including a time delay in communications after two months of simulated travel. The crew of 6 will simulate all aspects of a journey to the Red Planet, with a 250-day outward trip, a 30-day stay on its surface, and a 240-day return flight.

This will be an experiment like no other. The "crew" will live in a small steel container made up of four interlocked modules measuring, in total, 550 cubic meters to represent the interplanetary spacecraft. For 18 months, the six people will live inside this windowless environment, attempting to simulate the conditions onboard a spacecraft on a round-trip to Mars. During the experiment, researchers will assess the effect that isolation has on various psychological and physiological aspects such as stress, hormone levels, sleep quality, mood and the benefits of dietary supplements.

Of course no experiments on the ground can simulate the biological effect on the astronauts going to Mars for a long time. For example, NASA scientists have concluded that zero-gravity could hinder the growth of the human body. They reported that, spending time in zero-g may cause human cells to reproduce at slower rate. On the ground, gravity activates signals that tell cells to divide, so without gravity humans may suffer stunted growth. If these results are correct, the problem needs to be resolved before we consider long term life in space.

In the future, space exploration activities will be more complex and the trips will be longer to survey more distant and unknown localities. Systems of protection against space radiation, and progresses in space medicine could make interplanetary missions less severe. But to establish the presence prolonged in other worlds different from ours, many technological advances are necessary to be able to adapt the human body to extraterrestrial conditions.

Spaceships that include artificial gravity could help future space travelers ward off the debilitating loss of muscle and bone due to weightlessness on long missions. Astronauts could make the trip in a state of "forced hibernation". This it is the state that results from body freezing, in which the sudden suspension of the chemical reactions in the body due to the lack of oxygen puts the body in suspended animation. As Earth studies have revealed, frozen human beings can be revived.

The exploration of the Universe is the ultimate adventure, bursting with thrills and unforeseen experiences. And because it requires living in hostile environments, space exploration has always been and will continue to be risky, full of challenges and enormous danger. It does not matter the extent of their missions the space explorers will depart knowing that space is terribly dangerous. The brave men and women who join any space expedition do so knowing of the tremendous consequences involved. The strength of their desire to explore the space beyond the Earth must outweigh the risks they expect to face. Are you willing to join a space expedition?

★ ★ ★ ★ ★

Play Moonbase Apha, NASA's free new computer game and experience first hand what it would be to be a space explorer on a lunar base. Download Moonbase Alpha and team with your friends to play this videogame, and in a virtual environment, cooperatively race the clock to tackle the type of problems that future lunar astronauts will face. Access the game through the link provided on the *Resources* page.

13
Space Beckons Us!

The stars of our ancestors watch over us and call us

The phenomenal revolution sparked by humanity's reach of space incites us to raise our eyes to the stars with awe and reverence and also with longing. The first launch of an artificial satellite to Earth orbit in the twentieth century gave birth to the Space Age. Since then new astronomical discoveries are announced almost every day, amazing us with new facts that challenge our understanding. For the last twenty years the Hubble Space Telescope has given us spectacular and breathtaking views of the deep sky, revealing the ethereal and exquisite beauty of the Universe. What new epoch awaits us now that we begin a new program of human space exploration to reach farther frontiers in the Universe?

We live in the midst of a most extraordinary era of space discoveries of unprecedented magnitude, revealing every day a bit more of the marvelous creation of God. Many scientists, engineers and space technologists all over the world are coming together to develop launch vehicles, interplanetary spacecraft and space telescopes that are unraveling many mysteries of the cosmos. We continue pushing the limits of technology to design and develop the space ships that will take human beings to the Moon, to the asteroids, to Mars and beyond.

We have learned so much stimulated by our desire to reach the stars. Adding space observatories and automated spacecraft to the network of telescopes on the ground has opened more windows to the heavens. Just eighty years ago we didn't know how big the Universe really is. Now we have mapped its evolution and have seen the birth of stars. Just ten years ago astronomers began the search for planets circling around distant stars. Today, almost five hundred exoplanets have been detected, and many more wait behind the shinning of their stars to be exposed.

In fifty years we have vastly expanded our sense of space and determined clearly our place in the Universe. We have learned that we are not lost in the infinite cosmic sea. That with our star the Sun we are members of a beautiful and majestic galaxy that the ancient astronomers named the Milky Way. With every new discovery, space beckons us. The stars blink brightly and call us, challenging us all to discover their secrets.

Let's return to the original question. Can we make an interstellar voyage? Alas, human beings cannot navigate among the stars. Not now. Maybe one day in the distant future, perhaps in a few hundred years. For now we realize the distance between stars is enormous, incomprehensible large to make it across within the frame of reference of our human life span, even flying at the speed of light. Today, interstellar travel would take hundreds of years—the travel time to the neighborhood of the nearest star beyond the Sun—using our fastest spaceship that we could build. It took Voyager thirty years to reach the outer regions of our Solar System. It would have to travel for another 73,000 years to arrive near the closest stars.

Compare the highest escape velocity from Earth—about 16 km/s for the space probe New Horizons— with a spaceflight speed of just ten percent the speed of light ($v = 0.1c$). We quickly determine that to make it to the nearest stars we'd have to invent a relativistic rocket that would move at a velocity that is 4 orders of magnitude higher than the fastest spacecraft developed to date.

Nevertheless, we are capable of making interplanetary trips. To go to the closest planets in our Solar System we must accelerate scientific research and we must continue vigorously developing the required technologies to perfect the systems of propulsion. I believe that we are skillful and clever enough to extend our presence in the sky, and we should attempt in this century to go at least to Mars.

Seventy years ago many doubted that human beings could make a trip and land on the Moon. But humans did, and for the past five decades, advances in rocket technology enabled our space vehicles to break free of the grip of Earth's gravity and finally the first men visited the Moon. Since then, many people have been privileged to travel in space and live in space stations that orbit the Earth in a synchronous waltz that represents one of the greatest achievements of human ingenuity.

A human trip to another planet involves many challenges too difficult to quantify in this book. However, I believe human voyages to the Moon, to the asteroids and later to Mars are attainable within the next few decades, hopefully within my life time. First, however, people must return to the Moon. There is so much to do there to prepare humanity for reaching the next space frontier.

Of course we should go for a sustained human presence in space. Rather than visit the Moon for hours or a few days, the future space explorers must embark on missions that could last months. They will need new tools and technologies for living on the new worlds. Besides the challenge of designing these systems, engineers must build the potent vehicles than can transport all the extra supplies.

Aside from sample-return automated spacecraft, most robots in space missions go away holding a one-way ticket—they are not coming back! But for any human mission, we need a round-trip ticket. The voyage of exploration to Mars will be a much more complex endeavor, both in scale and in distance than any robotic or human space trip achieved so far. Yet, it is doable. We just have to mature and advance the appropriate technology.

Chemical rockets will continue to power the launch of spacecraft from Earth because chemical rockets provide the most lift thrust per weight capability. Then again, for crossing the vast distances between the planets in the Solar System, we need a different type of propulsion that can achieve higher vehicle velocities and use less propellant.

To reach any interplanetary destination beyond the Moon we have considered propulsion concepts that use other types of energy, hoping to significantly reduce trip time and the initial mass of the vehicle. Plasma rockets and thermo-nuclear rockets are two potential candidates that could be realized in the next few decades.

Of course, we also need to mature many other technologies, including life support systems, in order to build a lunar base so that we learn how to live in another world and eventually colonize the Red Planet. It won't be easy and it won't happen overnight. Maybe it will take hundreds of years for humanity to disperse and have our descendants born in other planets. However, the germination of ideas to realize space flight started with the bold imagination of a few bright, intelligent individuals. Now we need a new crop of bright, well-educated engineers and scientists to make possible our wildest dreams of deep space exploration, because to develop new ideas for crewed interplanetary travel, we must think beyond what we know today.

There was a time, during the first years of the space race, when the Moon must have seen as the farthest destination in the sky, the most difficult goal that humanity could aspire to reach. Today, we think the Moon is not such a great challenge and we believe instead that human beings must intent to dominate the cosmic dangers that stalk the Earth, such as the killer asteroids that move in a collision path with our planet. That is why the new space mission is considered to be a trip to an asteroid, one of the large objects near the Earth.

To land and set foot on the surface of the Moon for the first time was the most splendorous technical achievement for the people of the twentieth century, but now to conquer other challenges even more rigorous are the next goal, what defines the future of space exploration. For instance, if astronauts can make the trip and land on an asteroid, that means that humans are capable of destroying one or change its trajectory upon discovering that the rocky object is moving dangerously towards the Earth. Also, in addition to planning crewed missions, we could also build nuclear reactors in space to produce the necessary energy for rockets that could go beyond the asteroids.

In this century I hope that common citizens can travel to Earth orbit, go on vacations to the Moon or even live in another world in the sky. We are capable of establishing research settlements on the Moon, and I hope humans attempt many other activities of exploration in the sky to put in practice may ideas that we have already imagined. However, before space can be habitable, we must resolve many challenges. Engineering will develop new and better rocket propulsion systems. Space medicine will establish the treatments or protocols to overcome the physiological effects in the human body, and space psychology will help astronauts cope with the isolation and other mental health issues while traveling in long interplanetary routes.

Human beings, whose sense of distance may be defined by the unbounded view of the sky, may very well wonder how far we can go, how far we can really reach out if we build a spaceship to the stars. Some may feel bewildered and think that we cannot build a vehicle fast enough to go far enough within the Galaxy. Others may say that the light speed limit may stop our quest.

I believe that the exploration of space inevitably will lead us to conceive new ideas. Just trying something new or different will stretch our imagination. The limit will not be known until we try new ways and invent new technologies. In less than a hundred years humans applied the laws of physics that govern space travel, developed spacecraft to overcome the grip of Earth's gravity, sojourned in space and even walked on the Moon! So, I am sure than in a thousand years my descendants and your descendants will discover new ways to tour the cosmos by developing the technology to move from star to star, not just from planet to planet.

The future of space exploration offers the visionaries of the world many new possibilities. This is, after all, the essence of the quest, the calling of our souls to reach the heavens. As we gain more insight into the principles that govern the Universe, after we uncover more of its secrets, new ideas will germinate in our minds. As we learn more and are receptive to different possibilities, new phenomena as yet unimaginable may open new doors to the unknown of today.

The exploration of space has, and will continue to require, the best efforts and intellectual talents the world has available to make it reality. Remember, the Universe is indeed full of riddles and mysteries to be unraveled. Every day new discoveries uncover more secrets, and maybe you will be part of the new discoveries. I hope you are among those who resolve the technical challenges.

I hope you enjoyed reading about the scientific concepts and facts of the Universe interwoven in the *Prelude of Kuxan Suum*. I also wish these ideas will prompt you to take the first step (or accelerate it if you have already started) on the road of discovery. I hope you do because, in the end, we can no longer ignore our dreams and yearnings to explore the cosmos.

The stars of our ancestors watch over us and beckon us. You and I must continue this quest and go, like the Mayan princess, to seek the truth for our existence in this world, so that we may understand our role in the scheme of our vast and magnificent Universe. *Ad Astra*!

Questions to Ponder

1. If the princess Da'Lau were a beam of light, how long would it take her to cross the Solar System?
2. As a beam of light, how many light-years would it take her to cross the Milky Way Galaxy? A light-year is the distance that light can travel in one year — about 6 trillion miles (10 trillion kilometers).
3. What do you think happened to the princess when she fell into the black hole? Did she find the door to another universe, the heavens she longed for?
4. How far from us are the stars? How far is Proxima Centauri?
5. Could Da'Lau cross the Universe at speeds higher than the speed of light? If so, how?
6. Do you think people will be able to travel (some day) to other solar systems in our own Galaxy?
7. If astronomers discovered a planet like Earth orbiting a star located 10 light-years away, would humans be able to visit it? How long would it take for a spaceship to make the trip moving at half the speed of light?
8. The Milky Way galaxy is about 100,000 light-years across. Assuming that the spaceship moves at 10 percent of the speed of light, how long would it take for a spaceship to cross it?
9. What is "time"? Is time eternal?
10. In the fantasy, as Da'Lau became one with the stars, *past and future were interchanged.* What does this statement mean?
11. What is spacetime?
12. If the Universe began with the Big Bang, what created the cosmic egg? How did it get the initial energy to expand?
13. How many stars are in the visible Universe? How many galaxies are there?
14. Does the Universe have a center? Does it have an outer edge?
15. How big is the Solar System compared with the size of the known or visible Universe?
16. What makes the stars and planets move?
17. How do stars form? Why do stars die?
18. Will our Sun die, too? If so, how? When? What would happen to our Earth?
19. How far is the Sun from the center of the Milky Way?

20. If there are beings living on planets similar to Earth circling other distant stars, how would they look like?
21. Roughly 25 percent of the stuff in the Universe is called "dark matter" because it does not emit light or any other known radiation. How would you know dark matter is there if you cannot see it? How would you measure "dark energy"?
22. What are black holes and how do we know they exist?
23. What is the escape speed to overcome the gravitational pull of a black hole?
24. Are interstellar and intergalactic voyages possible for human beings? Explain.
25. Could people travel at the speed of light? Explain.
26. If you could travel in a spaceship moving at the speed of light, how would you slow down to land on an Earth-like planet?
27. What's the fastest spaceship humans have ever built? How does its speed compare with the speed of light?
28. How long would it take to travel to Proxima Centauri aboard the fastest spaceship built so far?
29. What is gravity? Is there gravity in outer space?
30. Do you think the Universe will continue expanding forever? If not, what do you think could happen?
31. If you could travel across the planets, where would you go? What would you do there? How would you live there? How many people you'd like to travel with you? Why?
32. What do you is more important, a trip to the Moon or a trip to the asteroids?
33. If astronauts arrive at one large asteroid, how would they land on it?
34. If you were a rocket scientist designing a space vehicle to take astronauts on a mission to Mars, how fast you'd like it to be? What type of rocket engines would you choose? Why?
35. What do you think of the idea to send astronauts to explore Jupiter's moons? How long would a round-trip last?

There are many more questions one can ask. I am sure you have other questions. Much of the work going on in astronautics, astrophysics, and cosmology focuses on those fundamental issues. And although it was not possible to address in depth these subjects in this book, I hope you find the answers.

Glossary

AEROBRAKING. A maneuver whereby a spacecraft uses the atmosphere of a planet to slow down and change its orbit.

ALPHA CENTAURI. Alpha Centauri is the 3-star system that is closest to the Earth. The dimmest star in the system, Proxima Centauri (Alpha Centauri C), is the closest star to us (other than our Sun). The stars Alpha Centauri A and Alpha Centauri B are close binary stars.

ANDROMEDA GALAXY. The Andromeda Galaxy (also known as M31 and NGC 224) is the closest major galaxy. It is a spiral galaxy (like our galaxy) and is in the Local Group. It can be seen with the naked eye in the constellation Andromeda.

ANTIMATTER. Matter made of particles with identical mass and spin as those of ordinary matter, but with opposite charge. Antimatter has been produced experimentally, but little of it is found in nature. Why this should be so is one of the questions that must be answered by any adequate theory of the early universe.

ANTIPARTICLE. Corresponding to most kinds of particles, there is an associated antiparticle with the same mass and opposite charges. (The exceptions are massless gauge bosons such as the photon.) Even electrically neutral particles, such as the neutron, are not identical to their antiparticle. In the example of the neutron, the *ordinary* particle is made out of quarks and the antiparticle out of antiquarks. The laws of nature were thought to be symmetric between particles and antiparticles until some experiments found that time-reversal symmetry is violated in nature. This small asymmetry is involved in baryogenesis, the process by which our universe came to consist almost entirely of matter, with almost no free antimatter.

ASTRONAUTICS. Astronautics is the science of space travel. Astronautical Engineering is the branch of engineering that deals with spacecraft designed to move or work entirely beyond the Earth's atmosphere. Astronautical Engineering is multidisciplinary and is also known as the science and technology of space flight.

Astronautics includes the research, design and development of space vehicles, and all other supporting technologies. From the moment a space mission is considered, whether to orbit the Earth or to travel beyond, we take into account the force of gravity. All space vehicles are subject to this force. The space age began with the launch of Sputnik 1 by the former Soviet Union on October 1957, followed with the launch of Explorer 1 by the United States on January 1958.

ATMOSPHERE. The atmosphere is the mixture of gases that surrounds a planetary object, moon, or star. The Earth's atmosphere is mostly nitrogen; the Sun's atmosphere is mostly hydrogen gas.

AURORA. Auroras are beautiful undulating sheets of light in the near-polar sky. They are caused by gases that become excited after being hit by solar particles. Most auroras are 100 to 250 km above the ground.

BLACK HOLE. A black hole is a massive object (or region) in space that is so dense that within a certain radius (the Schwarzschild radius, which determines the event horizon), its gravitational field does not let anything escape from it, not even light. It is thought that giant stars (those with a mass over 3 times the mass of the Sun) will evolve into red supergiants, then supernova, and then black holes. It is thought that the typical black hole has a mass of roughly 10 times that of the Sun, but the range must be huge. For a typical black hole with a mass 10 times that of the Sun, the Schwarzschild radius would be roughly 30 km (18.6 miles). The term *black hole* was coined by the physicist John Archibald Wheeler. Astronomers think that there is a black hole at the center of each galaxy.

CHEMICAL ROCKET. An engine in which one, two or more propellants are mixed together in a high-pressure chamber to produce a powerful chemical reaction. The reaction produces hot gases that are forced out a nozzle at high speed, creating thrust and propelling the rocket forward. There are two types, solid and liquid propellant rockets. The Space Shuttle uses two solid rocket boosters, and three liquid propellant main engines. All launch vehicles use solid rocket propulsion to boost a spacecraft to orbit.

COMET. A comet is a small, icy celestial body that orbits around the Sun. It is made up of a nucleus (solid, frozen ice, gas and dust), a gaseous coma (water vapor, CO2, and other gases), and a long tail (made of dust and ionized gases). The tail develops when the comet is near the Sun. The comet's long ion tail always points away from the Sun, because of the force of the solar wind. The tail can be up to 250 million km long, and is most of what we see. Comets are only visible when they're near the Sun in their highly eccentric orbits.

COSMIC MICROWAVE BACKGROUND. This is a form of electromagnetic radiation that fills the entire universe. The Big Bang theory predicts that the early universe was a very hot place and that as it expands, the gas within it cools. Thus the Universe should be filled with radiation that is literally the remnant heat left over from the Big Bang, called the "cosmic microwave background radiation", or CMB. Discovered in 1965, this radiation is considered to be the best evidence for the Big Bang model of the Universe.

DARK MATTER. Dark matter is unknown matter that may constitute as much as 22 percent of the matter in the Universe.

DARK ENERGY. A hypothetical type of energy that fills the empty space and tends to increase the rate of expansion of the Universe. Most recent observations suggest the Universe made up of 74% dark energy, 22% dark matter, and 4% ordinary matter.

DE-ORBIT BURN. The firing of a spacecraft's engine against the direction of motion to reduce its orbital speed. The speed reduction places the spacecraft in a lower orbit. If this lower orbit passes through Earth's atmosphere, the spacecraft reenters.

DWARF STAR. Small star composed mostly of electron-degenerate matter. White dwarfs have mass comparable to the Sun's and their volume is comparable to the Earth's, they are very dense.

EARTH-MARS DISTANCE. The distance between the two planets varies from 56 million km to over 400 million km (35 million miles to 250 million miles). This is how far we want to go after the exploration of the Moon.

ESCAPE VELOCITY. The escape velocity is how fast an object has to move away from a planetary object in order to escape its gravitational field. The outward velocity required to leave the surface of a body with mass M and radius R and escape to infinity (not fall back). The formula for the escape velocity is $(2GM/R)^{1/2}$.

ELLIPTICAL ORBIT. An orbit is the path in space one object takes about another. On an elliptical orbit the path defines the shape of an ellipse (a regular oval).

EVENT HORIZON. The event horizon is the radius from a black hole inside of which it is impossible to escape (a "point of no return" called the Schwarzschild radius). It is also the radius at which a mass must be compressed down to in order to turn it into a black hole.

FLYBY. The motion of a spacecraft when it goes past a planet or moon (the vehicle neither orbits the planet nor it lands on it).

GALAXY. A huge group of stars and other celestial bodies bound together by gravitational forces. There are spiral, elliptical, and irregularly shaped galaxies. Our Sun and entire Solar System are a small part of the Milky Way Galaxy.

GAMMA-RAYS. The most energetic form of electromagnetic radiation followed, with progressively lower energies, by X-rays, ultraviolet rays, optical radiation (light), infrared radiation, and radio waves.

GRAVITY. Gravity is a physical force that pulls objects together. Every bit of mass produces a gravitational force; this force attracts all other masses. The more massive an object, the stronger the gravitational force. Newton formulated the laws of gravity.

GRAVITATION (force in Newton's law of universal gravitation). The gravitational attraction between bodies with mass. It is a part of classical mechanics first formulated by Newton in his book the *Principia.* The law of universal gravitation states that: "Every point mass attracts every other point mass by a force pointing along the line intersecting both points." The force is proportional to the product of the two masses and inversely proportional to the square of the distance between the point masses: $F = G\frac{m_1 m_2}{r^2}$, where F is the magnitude of the gravitational force between the two point masses, G is the gravitational constant, m_1 is the mass of the first point mass, m_2 is the mass of the second point mass, and r is the distance between the two point masses. In SI units, F is measured in newtons (N), m_1 and m_2 in kilograms (kg), r in meters (m). G is equal to 6.67×10^{-11} N m^2 kg^{-2}.

GRAVITY-ASSIST. A maneuver whereby a spacecraft flies past a planet and takes up a small fraction of the planet's orbital energy. This additional energy allows the space vehicle to change direction and speed. The gravity-assist maneuver reduces significantly the amount of propellant a vehicle needs to carry. It has been used successfully in many interplanetary missions such as the *Cassini-Huygens*, which used gravity-assist from Venus, Earth, and Jupiter to boost its trip to Saturn.

HEAVY LIFT LAUNCH VEHICLE (HLLV). A launch vehicle capable of lifting more mass into Low Earth Orbit (LEO) than Medium Lift Launch Vehicles. While there is no universally accepted capability requirement, rockets like the Saturn V, Titan IV, Ariane 5, Proton, and Delta IV-Heavy are generally considered Heavy Lift launch vehicles. HLLVs are the only rocket-powered launchers capable of moving heavier satellites into geostationary or geosynchronous orbit. The capability of achieving geostationary transfer orbit is critical to the placement of modern satellites, as well as to the success of future space programs going to the Moon and to Mars

HELIOSPHERE. It is an enormous magnetic bubble around the Sun. The bubble is formed by the solar wind, which is made up of high energy particles shooting out from the Sun and push out the interstellar medium coming in from beyond the Solar System.

INTERSTELLAR DUST. Interstellar dust is composed of microscopic bits (on the order of a micron in diameter) of carbon and/or silicates. The origin of interstellar dust in unknown, but it seems to be associated with young stars.

LIGHT. Light is a type of energy (and the tiny part of the electromagnetic spectrum that we can see). The fastest that light can travel is approximately 300,000 km/s (186,300 miles per second) in a vacuum. The dimension of the Solar System can be inferred from the time it takes for a ray of light to cross it, which is about 11 light-hours.

LOW EARTH ORBIT (LEO). An orbit within the locus extending from the Earth's surface up to an altitude of 2,000 km. Most satellites orbit near the top of the Earth's atmosphere, about 160-320 km. These satellites, including the International Space Station, orbit the Earth in about 90 minutes.

MAIN SEQUENCE. A curve in the Hertzsprung-Russell (H-R) diagram, used to explain the evolution of stars. A star's position in the H-R diagram represents its brightness and its temperature. Stars on the left of the diagram are blue because they're hotter, and stars on the right are red because they're cooler. The diagonal band from the upper left corner to the lower right corner is called "main sequence." The diagram was developed independently by the Danish astronomer Ejnar Hertzsprung, and by the American Henry Norris Russell.

MAIN SEQUENCE STAR. A star that derives its energy from the conversion of hydrogen into helium in its core. The Sun is a main sequence star.

MICROQUASAR. A high-energy binary star system that includes a black hole or neutron star and which resembles a quasar.

MILKY WAY GALAXY. The Milky Way Galaxy is a spiral galaxy; our Sun, or better yet, our Solar System is a small part of it. Most of the stars that we can see are in the Milky Way Galaxy. The main plane of the Milky Way looks like a faint band of white in the night sky. The Milky Way is about 100,000 light-years in diameter and 1,000 light-years thick. There are about 2×10^{11} stars in the Milky Way.

This spiral galaxy formed about 14 billion years ago. It takes the Sun roughly 250 million years to orbit once around the Milky Way. The Earth is about 26,000 light-years from the center of the Milky Way Galaxy.

NEBULA. A nebula is a huge, diffuse cloud of gas and dust in intergalactic space. The gas in nebulae (the plural of nebula) is mostly hydrogen gas (H_2).

OMEGA NEBULA. The Omega Nebula Messier 17 (M17, NGC 6618), also called the Swan Nebula, the Horseshoe Nebula, or (especially on the southern hemisphere) the Lobster Nebula, is a region of star formation and shines by excited emission, caused by the higher energy radiation of young stars. Unlike in many other emission nebulae, however, these stars are not obvious in optical images, but hidden in the nebula. Star formation is either still active in this nebula or ceased very recently. A small cluster of about 35 bright but obscured stars seems to be imbedded in the nebulosity.

PLEIADES. A beautiful cluster of stars, which can be seen with the naked eye (we don't need binoculars or a telescope to see). We can see about seven stars in the cluster, for which they're known as the Seven Sisters. However, there are thousands of stars in this cluster, including cool faint brown dwarf stars. The distance to the Pleiades is estimated to be about 440 light-years.

POWERS OF TEN:

Ten	$10 = 10^1$
One hundred	$100 = 10^2$
One thousand	$1000 = 10^3$
One million	$1{,}000{,}000 = 10^6$
One billion	$1{,}000{,}000{,}000 = 10^9$
One trillion	$1{,}000{,}000{,}000{,}000 = 10^{12}$

A billion is a thousand million. The Universe is 13.75 billion years old (13,750 million year). The Earth formed about 4.54 billion years ago. Multi-cellular life evolved on Earth about a billion years ago. And 2.00×10^5 years ago since humans started looking like they do.

PROXIMA CENTAURI. It is a red dwarf star, the nearest star to the Sun, approximately 4.2 light-years away, in the constellation of Centaurus. It was discovered in 1915 by Robert Innes, the Director of the Union Observatory in South Africa. Proxima Centauri, also known as Alpha Centauri C, is the dimmest star in the Alpha Centauri system.

PULSAR. A rotating neutron star whose radiation is observed as regular pulses.

QUASAR. It is an extragalactic object similar to a star that is among the most luminous and thought to be the most distant objects in the Universe. Quasar is a word derived from the term Quasi Stellar Object.

ROCKET CHEMICAL PROPELLANT. A chemical propellant consists of a fuel and an oxidizer that produce a chemical reaction in the combustion chamber from which a hot gas at high velocity is expelled to create thrust. Rocket chemical propellants can be in solid or liquid form. Most launch vehicles use booster rockets that burn solid propellants.

RED GIANT. A post-main sequence star of modest mass (a few solar masses or less) with an extended, relatively cool atmosphere.

SINGULARITY. A point in spacetime at which the density of matter and the gravitational field are infinite (forming a black hole). Singularities are points at which the mathematical solutions to the spacetime equations are undefined.

SOLAR WIND. The flow of fast-moving energetic particles that escape from the Sun.

SPACE TRAVEL. Space travel is that achieved by a vehicle that leaves Earth's atmosphere. The science and engineering of space travel crewed or not, is called astronautics. The exploration of space or astronautics is an interdisciplinary science that is built upon the knowledge of other fields such as physics, astronomy, mathematics, chemistry, biology, medicine, electronics, and others. The science that studies space travel is also known as cosmonautics because travel occurs in the cosmos. The name astronautics is used mainly in the West; that is why the American spacecraft crew members are known as astronauts, while the space travelers from the former Soviet Union are known as cosmonauts. In 2003, two new terms were coined to refer to the space travelers from China: *yuhangyuans* or *taikonauts*. The word *yuhangyuan* means space navigator, while the word taikonaut is derived from *taikong*, the Chinese word for space.

STAR. A self-luminous gaseous body that typically generates energy by nuclear reactions in its interior. White dwarfs and neutron stars that no longer possess nuclear reactions, but shine by radiating stored-up heat that originally was derived from nuclear reactions, are also referred to as stars.

SUPERMASSIVE BLACK HOLE. A black hole that has a million or as much as a billion solar masses. Such huge black holes lurk at the centers of many active galaxies.

SUPERNOVA. The explosion of a star. Supernovae come in two types: Type I is caused by sudden nuclear burning in a white dwarf star. Type II is caused by the collapse of the core of a supermassive star at the end of its nuclear-burning life. In either case, the star is destroyed and the light given off in its explosion briefly rivals the total light given off by a whole galaxy.

SUPERNOVA REMNANT. The material blown off during a supernova, now seen as a great glowing cloud expanding into space.

TERMINATION SHOCK. It is the outer envelope of the heliosphere, beyond the orbit of Pluto, where the solar wind collides head-on with the interstellar medium, causing the particles that were traveling outwards from the Sun at a million miles an hour to abruptly slow down to subsonic speeds. This boundary is somewhere between 85 and 120 AUs from the Sun (13.5 billion km away).

TIME DILATATION. Concept that arises from Einstein's theories of relativity. It's the phenomenon whereby an observer finds that another's clock which is physically identical to their own is ticking at a slower rate as measured by their own clock. This is often taken to mean that time has "slowed down" for the other clock, but that is only true in the context of the observer's frame of reference.

WARP DRIVE. A form of faster-than-light propulsion system, capable of propelling spacecraft or other bodies to speeds many times the speed of light, while avoiding the problems associated with time dilation.

WHITE DWARF. The remnant of a star, at the end of its life, consisting of a carbon and oxygen core supported by electron degeneracy pressure. The surface has a very high temperature and radiates mainly in the ultraviolet (hence white as in white hot), but it is only about the size of the Earth (hence dwarf).

WORMHOLE. A wormhole in space (also known as Einstein-Rosen Bridge, named for Albert Einstein and Nathan Rosen) is a mathematical solution to Einstein's theory of general relativity. A wormhole would theoretically provide a shortcut through widely-separated parts of spacetime, through a black hole and out of a white hole (moving faster than the speed of light).

Other Resources

American Institute of Aeronautics and Astronautics, AIAA
www.aiaa.org
AstroMía
http://www.astromia.com/
Books and Theses on ExtraSolar Planets and Exobiology
http://exoplanet.eu/biblio-bookThese.php
Coalition for Space Exploration
http://spacecoalition.com/
Encyclopedia of the Cosmos
www.eofcosmos.org
ESA Mars 500 - Overview of the human spaceflight program to study human endurance on a simulated trip to Mars
http://www.esa.int/esaMI/Mars500/
Heavens Above
www.heavens-above.com
Hubble Space Telescope Site
http://hubblesite.org
Jet Propulsion Lab
www.jpl.nasa.gov
Kepler Mission
http://kepler.nasa.gov
Kuxan Suum
www.kuxansuum.net
Millis, Mark. Annotated Bibliography (space warps)
www.nasa.gov/centers/glenn/research/warp/bibliog.html
NASA
www.nasa.gov
NASA's Moonbase Alpha – Virtual Lunar Life PC game free download
http://store.steampowered.com/app/39000/
Space Weather
www.spaceweather.com
The Planetary Society
http://planetary.org

Dora Musielak holds a Ph.D. degree in aerospace engineering, specialized in rocket propulsion. She has spent her career conducting research in the aerospace industry and as a professor teaching science mathematics and engineering courses at several universities, most recently at the University of Texas in Arlington. Dr. Musielak is the recipient of two NASA research fellowships, and she is an Associate Fellow of the American Institute of Aeronautics and Astronautics (AIAA).

Dr. Musielak is an enthusiastic promoter of the history of women in science and mathematics and has given numerous talks on the topic. She is the author of *Sophie's Diary* (2004) which, through a fictionalized journal, tells the story of a young mathematician in Paris between 1789 and 1793. The story was inspired by the life of French mathematician Sophie Germain who made important contributions to number theory and mathematical physics. This book was translated into Spanish as *El Diario de Sofi.*

CPSIA information can be obtained at www.ICGtesting.com
Printed in the USA
BVOW08s0319180515

400778BV00001B/44/P